NIOSH
The National Institute for Occupational Safety and Health

Delivering on the Nation's Investment in Worker Safety and Health

It has been nearly 40 years since the Occupational Safety and Health Act of 1970 was passed. During that time, NIOSH has worked diligently to ensure that U.S. workers are safe from occupational illness, injuries, and fatalities while at work. Our strong scientific foundation has guided our work as we strive to fulfill the responsibilities of the Act, and to carry out the duties entrusted to us by Congress.

NIOSH's research and recommendations over the years have made a significant impact in reducing and preventing occupational injuries, illnesses, and fatalities. Our work has lead to recommendations on reducing exposures to asbestos, lead, vinyl chloride, and other toxic industrial agents. As the U.S. economy has changed NIOSH has kept pace by addressing the new occupational hazards that have arisen or become more prominent, such as latex allergies, musculoskeletal disorders, indoor air quality, and workplace violence. And with the goal of achieving even greater impact with our research, NIOSH created the National Occupational Research Agenda (NORA) in 1996. The creation of NORA allowed us to expand our partnerships and leverage resources to meet the needs and challenges of the changing face of work.

This document provides a snapshot of our work addressing the safety and health issues that reach across all the U.S. states, industries, and disciplines. Here we have included information about our efforts in traditional and emerging areas such as NORA, research-to-practice, emergency response, nanotechnology, personal protective technology, global collaborations, and other cross-cutting programs. We have also included examples of how NIOSH and our partners are working hard to achieve our shared mission of making the workplace safer and healthier for all workers.

As we enter the second decade of the 21st Century, the face of the U.S. economy and the challenges and risks workers face continues to change. These challenges may range from assisting injured, returning military personnel in making the transition to safe, fulfilling civilian work, to the potentially unknown hazards to workers employed in the discovery, development, and production of new sources of renewable energy. Since 1970 we have developed new knowledge and new scientific techniques that can be applied to the workplace. As we continue to move forward, we must also look back to see where we can make changes based on this knowledge to continue to improve the safety and health of all workers. These are just a few of the challenges that NIOSH and our nation face as we look towards the future.

As the U.S. looks to the workplace of tomorrow and how we will maintain leadership in the global market, it is important that the safety and health of workers is made an integral part of that strategy. I hope you find this document interesting and engaging and that it stimulates new ideas for ways in which we might collaborate to protect our nation's workers.

John Howard, MD
Director, National Institute for Occupational Safety and Health

About NIOSH	1
The Current State of Worker Safety & Health	7
Introduction to NORA	11
Agriculture, Forestry, & Fishing	14
Construction	16
Healthcare & Social Assistance	18
Manufacturing	20
Mining	22
Services	24
Transportation, Warehousing, & Utilities	26
Wholesale & Retail Trade	28
Partnering with NIOSH	31
Introduction to NIOSH Cross-Sectors	35
Authoritative Recommendations	36
Cancer, Reproductive, & Cardiovascular Diseases	36
Communication & Information Dissemination	37
Economics	38
Emergency Preparedness & Response	38
Engineering Controls	39
Exposure Assessment	39
Global Collaborations	40
Health Hazard Evaluations	40
Hearing Loss	41
Immune & Dermal Diseases	42
Musculoskeletal Disorders	43
Nanotechnology	43
Occupational Health Disparities	44
Personal Protective Technologies	44
Prevention Through Design	45
Radiation Dose Evaluation	46
Respiratory Diseases	47
Small Business Assistance & Outreach	47
Surveillance	48
Training Grants	50
Traumatic Injury	50
WorkLife	51
Work Organization & Stress-Related Disorders	52
Challenges About Nanotechnology	55
National Academies' Review	56
Research-To-Practice	59

Photo by David Parker

Two roughnecks work on the rig floor to drill a new oil well. These floor hands are attaching a new length of pipe to the drill.
Photo by Elaine Cullen

NIOSH (nahy-osh), n

The National Institute for Occupational Safety and Health

1. the only federal agency mandated to conduct research and make recommendations to prevent work-related injury, illness, and death; 2. a public health science organization that provides and funds research, information, education, and training in the field of occupational safety and health with an annual budget of approximately $300 million; 3. a federal government agency that is committed to the development and integration of knowledge and intervention and is dedicated to building and maintaining productive partnerships; 4. a world-renowned research institute that attracts highly talented investigators from many disciplines.

Function: To protect the more than 145 million U.S. workers through research and education.

Origin: Established under the Occupational Safety and Health Act of 1970.

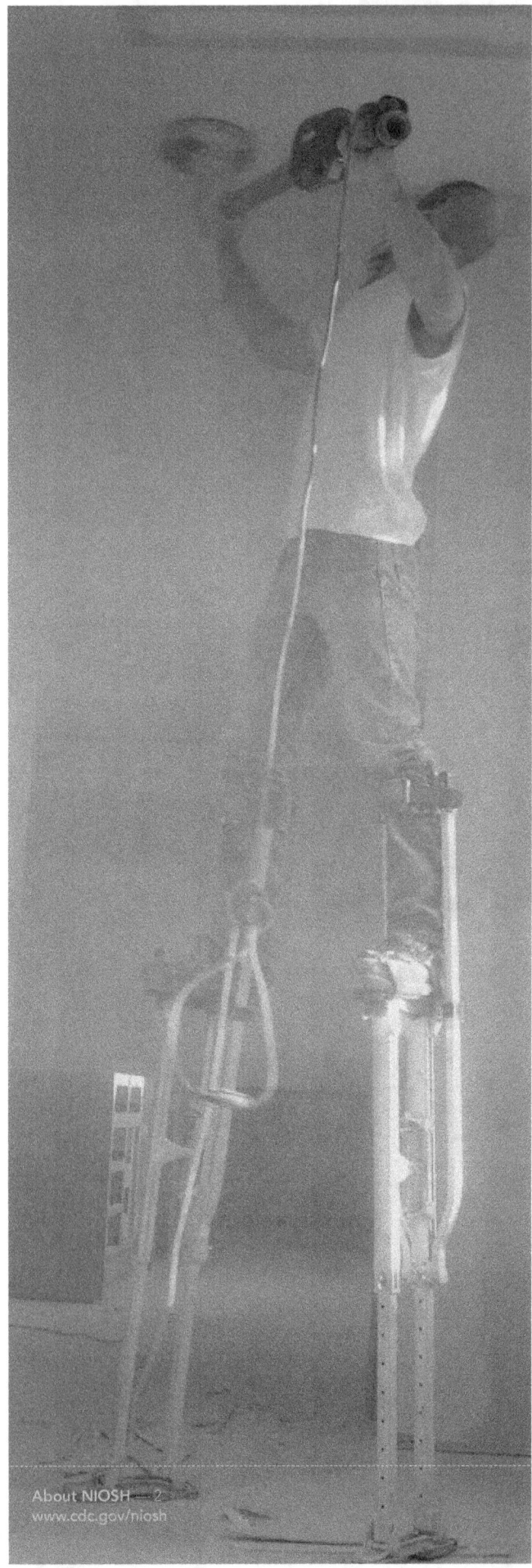

AFF	Agriculture, Forestry, & Fishing
AIDS	Acquired Immune Deficiency Syndrome
AIHA	American Industrial Hygiene Association
ASHP	American Society of Health-System Pharmacists
ASSE	American Society of Safety Engineers
ASTM	American Society for Testing & Materials
BLS	Bureau of Labor Statistics
CDC	Centers for Disease Control & Prevention
DOL	U.S. Department of Labor
EHOPAC	Environmental Health Officer Professional Advisory Committee
EPA	U.S. Environmental Protection Agency
FAA	Federal Aviation Administration
FLC	Federal Laboratory Consortium
HETAB	Health Evaluations & Technical Assistance Branch (within NIOSH)
HHS	U.S. Department of Health & Human Services
HIV	Human Immunodeficiency Virus
HSA	Healthcare & Social Assistance
IARC	International Agency for Research on Cancer
ILO	International Labour Organization
ISO	International Organization for Standardization
KOSHA	Korea Occupational Safety & Health Agency
MINER	Mine Improvement & New Emergency Response (Act of 2006)
MSHA	Mine Safety & Health Administration
NAICS	North American Industry Classification System
NASA	National Aeronautics & Space Administration
NHCA	National Hearing Conservation Association
NIH	National Institutes of Health
NIOSH	National Institute for Occupational Safety & Health
NORA	National Occupational Research Agenda
NSC	National Safety Council
OECD	Organisation for Economic Co-Operation & Development
OSH	Occupational Safety & Health
OSHA	Occupational Safety & Health Administration
PAHO	Pan American Health Organization
R2P	Research-to-Practice
TWU	Transportation, Warehousing, & Utilities
WHO	World Health Organization
WRT	Wholesale & Retail Trade
>	Greater Than
<	Less Than
≤	Less Than or Equal To
x	Times (when used in Relevant Information & impact narratives)
📖	Program Reviewed by the National Academies
👥	Narrative of Impact

Photo by mel0808johnson, downloaded from Flickr

NIOSH Mission & Responsibilities

The mission of NIOSH is to generate new knowledge in the field of occupational safety and health and to transfer that knowledge into workplace practice to prevent work-related injury, illness, and death. To accomplish this mission, NIOSH provides national and international leadership by conducting scientific research, identifying causes of work-related disease and injury, identifying the potential hazards of new work technologies and practices, creating ways to prevent workplace hazards, developing guidance and authoritative recommendations, and delivering products and services to its various stakeholders.

Section 22 of the Occupational Safety and Health Act outlines the specific responsibilities of NIOSH:

Develop recommendations for occupational safety and health standards,

Perform all functions of the Secretary of Health and Human Services under Sections 20 and 21 of the Act,

    Sec. 20: conduct research on worker safety and health

    Sec. 21: conduct training and employee education

Develop information on safe levels of exposure to toxic materials and harmful physical agents and substances,

Conduct research on new safety and health problems,

Conduct onsite investigations to determine the toxicity of materials used in workplaces, and

Fund research by other agencies or private organizations through grants, contracts, and other arrangements.

> With today's economic, healthcare, and workers' compensation crises challenging our country, prevention of work-related injury, illness, and death is critically important.

Distinguishing NIOSH from OSHA

The Occupational Safety and Health Act of 1970 created both the Occupational Safety and Health Administration (OSHA) and NIOSH. Although NIOSH and OSHA were established by the same Act of Congress, the two agencies have distinct and separate responsibilities.

OSHA is housed under the U.S. Department of Labor (DOL), whereas NIOSH is part of the Centers for Disease Control and Prevention (CDC) under the U.S. Department of Health and Human Services (HHS).

OSHA is responsible for setting and enforcing workplace standards and regulations, whereas NIOSH is responsible for helping assure safe and healthful working conditions through research, information, education, and training in the field of occupational safety and health.

OSHA approves and monitors state-based safety and health plans and provides up to 50 percent of an approved plan's operating costs, whereas NIOSH funds training grant programs throughout the country to develop qualified practitioners and researchers who will contribute to the prevention of work-related injury, illness, and death.

Aside from these distinctions, NIOSH and OSHA work collaboratively to assure the safety and health of every U.S. worker.

Serving the Nation Through Research & Education

NIOSH is actively engaged in and committed to training tomorrow's leaders in occupational safety and health. The information below depicts locations of NIOSH laboratories; its Centers for Agricultural Disease and Injury Research, Education, and Prevention; Training Project Grants; Education and Research Centers; WorkLife Centers of Excellence; Liaison to CDC State & Local Support Office; and National Construction Center.

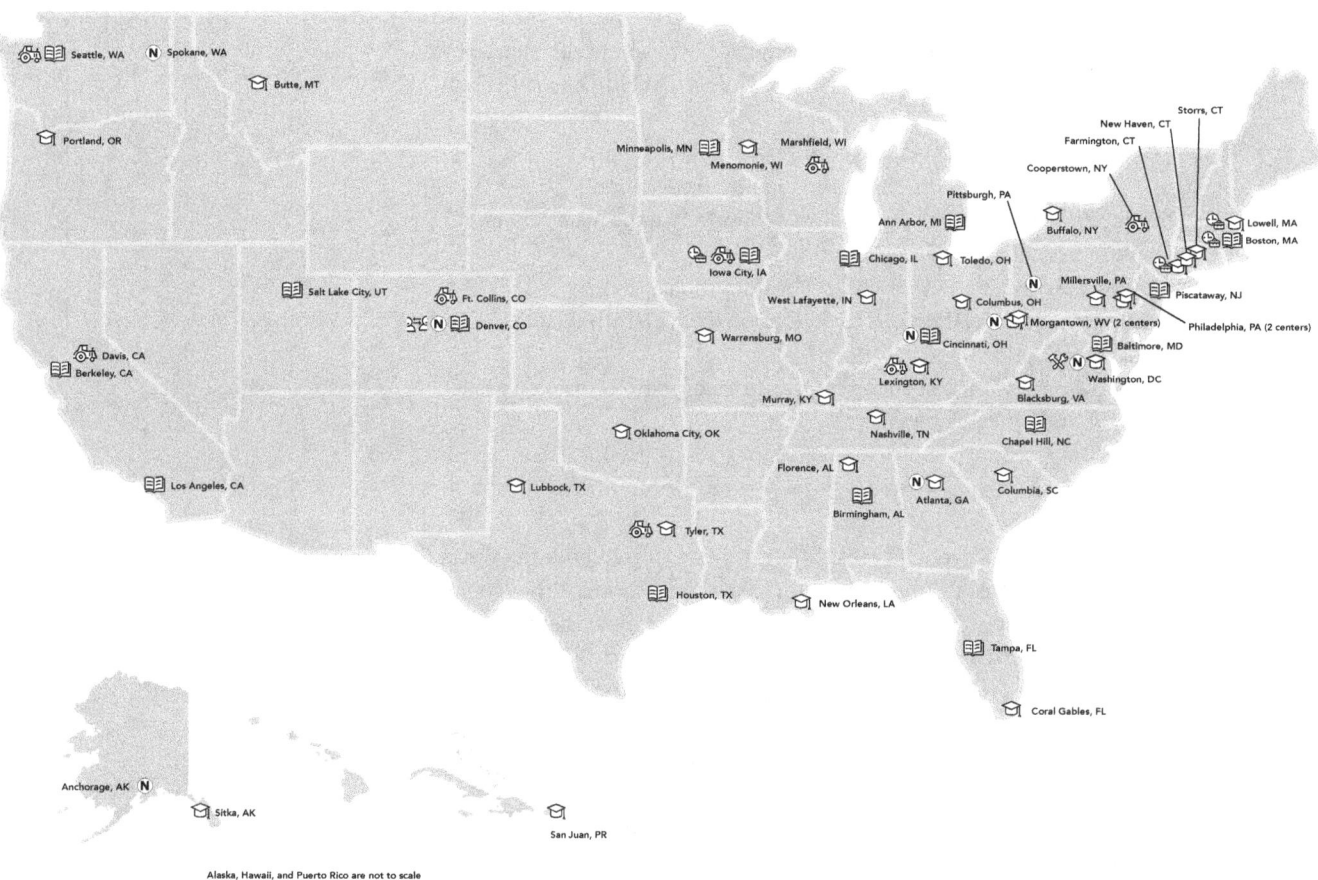

NIOSH Facilities

Facilities are headquartered in Washington, District of Columbia and Atlanta, Georgia with laboratories, offices, and staff in Anchorage, Alaska; Cincinnati, Ohio; Denver, Colorado; Morgantown, West Virginia; Pittsburgh, Pennsylvania; and Spokane, Washington.

Centers for Agricultural Disease & Injury Research, Education, & Prevention

Centers represent a major effort to protect the safety and health of agricultural workers and their families. These Centers of Excellence were established in 1990 as part of an Agricultural Safety and Health Initiative and are strategically located to be responsive to issues unique to the different regions across the country. The centers were established by cooperative agreement to conduct research, education, and prevention projects to address the nation's pressing agricultural safety and health problems. NIOSH currently supports 8 centers across the country.

Training Project Grants

Training Project Grants are awarded to academic institutions that primarily have *single-discipline* graduate training programs in industrial hygiene, occupational health nursing, occupational medicine, occupational safety, and other closely-related fields. Training Project Grants provide academic and research training programs that develop professionals who will contribute to the prevention of work-related injury, illness, and death. NIOSH currently supports 32 Training Project Grants across the country. See Training Grants (page 50) for more information.

Education & Research Centers

Centers are located at academic institutions that emphasize *interdisciplinary* graduate training programs in industrial hygiene, occupational health nursing, occupational medicine, occupational safety, and other closely-related fields. Education and Research Centers strongly encourage collaboration among various occupational safety and health disciplines and provide academic and research training programs that develop professionals who will contribute to the prevention of work-related injury, illness, and death. NIOSH currently supports 17 centers across the country. See Training Grants (page 50) for more information.

WorkLife Centers of Excellence

Centers represent the NIOSH initiative to sustain and improve worker health through better work-based programs, policies, and practices. The overall health of workers is influenced by many factors inside and outside the workplace, such as stress, physical and chemical exposures, diet, exercise, smoking, medication, and alcohol use. NIOSH currently funds 3 centers located at 4 universities to support and expand multidisciplinary research, training, and education in the area of WorkLife. See WorkLife (page 51) for more information.

Liaison to CDC State & Local Support Office

NIOSH is strategically positioned to work with states to enhance occupational safety and health activities. The NIOSH state liaison works closely with the new CDC State and Local Support office serving as an advocate for NIOSH priorities and objectives that support, foster, and promote state-based occupational safety and health activities.

National Construction Center

The National Construction Center serves as an integrated, multi-disciplinary occupational safety and health resource for the entire U.S. Construction sector. NIOSH has funded a national construction center since 1994. In addition to performing research and prevention activities, the Center also serves to transfer research findings into construction practice via an Electronic Library of Construction Safety and Health (eLCOSH) and other initiatives. It also periodically publishes a Construction Chart Book to provide researchers and the industry with key statistics. The current National Construction Center recipient is CPWR—The Center for Construction Research and Training, based in Silver Spring, Maryland. The Center includes a university consortium with academic partners in 8 states.

Structural Organization of NIOSH
Strategically Located to Address Safety & Health

Office of the Director
- Alaska Regional Office
- Denver Regional Office
- Office of Administrative & Management Services
- Office of the Associate Director for Science
- Office of the Chief of Staff
- Office of Compensation, Analysis, & Support
- Office of Emergency Preparedness & Response
- Office of Extramural Programs
- Office of Health Communication & Global Collaboration
- Office of Minority Health
- National Occupational Research Agenda Coordination Office
- Office of Planning & Performance
- Office of Research & Technology Transfer

Office of Mine Safety & Health

Pittsburgh Research Laboratory
- Respiratory Hazards Control Branch
- Hearing Loss Prevention Branch
- Disaster Prevention & Response Branch
- Mining Injury Prevention Branch
- Surveillance & Research Support Branch
- Rock Safety Engineering Branch

Spokane Research Laboratory
- Extramural Coordination & Information Dissemination Activity
- Mining Surveillance & Statistics Support Activity
- Mining Injury & Disease Prevention Branch
- Catastrophic Failure Detection & Prevention Branch

Division of Applied Research & Technology
- Biomonitoring & Health Assessment Branch
- Chemical Exposure & Monitoring Branch
- Engineering & Physical Hazards Branch
- Organizational Science & Human Factors Branch

Division of Respiratory Disease Studies
- Field Studies Branch
- Laboratory Research Branch
- Surveillance Branch

Division of Surveillance, Hazard Evaluations, & Field Studies
- Hazard Evaluations & Technical Assistance Branch
- Industrywide Studies Branch
- Statistical Support Most Efficient Organization
- Surveillance Branch

Division of Safety Research
- Analysis & Field Evaluations Branch
- Protective Technology Branch
- Surveillance & Field Investigations Branch

Education & Information Division
- Document Development Branch
- Information Resources & Dissemination Branch
- Risk Evaluation Branch
- Training Research & Evaluation Branch

Health Effects Laboratory Division
- Allergy & Clinical Immunology Branch
- Biostatistics & Epidemiology Branch
- Engineering & Control Technology Branch
- Exposure Assessment Branch
- Pathology & Physiology Research Branch
- Toxicology & Molecular Biology Branch

National Personal Protective Technology Laboratory
- Technology Evaluation Branch
- Technology Research Branch
- Policy & Standards Development Branch

The Current State of Worker Safety & Health
Understanding the Dynamics of a Changing Workforce

In 2008 more than 145 million people in the U.S. were employed in the civilian workforce. Every day, approximately 9,000 workers are injured on the job and 15 workers die from a fatal workplace injury. Work-related illness claims the lives of about another 135 workers and retirees daily. According to the Bureau of Labor Statistics (BLS), 5,657 workers died from work-related injuries and more than 4 million nonfatal injuries and illnesses were reported in 2007. The economic impact of work-related injury and illness has been estimated to be $171 billion annually, the same as cancer or cardiovascular disease and much greater than the burden from HIV/AIDS or Alzheimer's disease. In 2006 employers spent an estimated $87.6 billion on wage payments and medical care for workers hurt on the job.

Addressing workplace safety and health poses numerous challenges. First, the composition of the U.S. workforce is becoming increasingly diverse; it is becoming older, more racially and ethnically diverse, and more women are entering the workforce. These changes reflect the changing social and demographic characteristics of the country but also produce new safety and health issues. It has become clear that certain populations experience an increased burden of disease, disability, and death. These populations also frequently have less access to quality healthcare.

Moreover, U.S. workplaces are rapidly evolving, changing the way work is organized. Jobs in our economy continue to shift from manufacturing to services, with service-providing industries now employing about 80 percent of all workers. Longer hours, compressed work weeks, shift work, reduced job security, and part-time and temporary work are realities of the modern workplace and are increasingly affecting the health and well-being of workers and their families. In addition, new chemicals, materials, processes, and equipment with new potential occupational risks are being developed and marketed at an ever-accelerating pace.

The acute and long-term effects of work-related injury, illness, and death translate into tremendous economic and emotional costs to society. Data show that when interventions, such as safe work practices and engineering controls, are based on sound scientific research the burden of injury and illness is significantly reduced. Through its Research-to-Practice (r2p) initiative, NIOSH works closely with its partners to move research findings and technologies out of the Institute and into the workplace, and to promote the diffusion of products and information in an effort to protect workers and reduce cost to employers, workers, their families, and society as a whole.

Injuries and illnesses become more severe with age. In 2007, workers aged 65 and older required a median of 16 days away from work after suffering a work-related injury or illness, whereas workers aged 16-24 years required only 4 days away from work.

In 2005, blue-collar jobs made up 23% of all jobs but these workers experienced 55% of all work-related injuries and earned an average of $15 per hour. White-collar workers earned more than double and experienced only 2% of the injuries on average.

In 2007, the death rate from work-related injuries for Hispanic workers was 4.6 per 100,000 workers, as compared with rates of 3.8 for all workers, 3.8 for non-Hispanic white workers, and 3.9 for non-Hispanic black workers.

More than 2/3 of Hispanic or Latino workers who suffered a fatal work-related injury in 2007 were born outside the U.S.

More than 50% of the 4 million work-related injury & illness cases reported in 2007 required days away from work, job transfer, or restriction.

In 2007, 92% of work-related fatalities occurred among men but men represented only 54% of the U.S. workforce.

Data used in this document were obtained from the following external sources: the Bureau of Justice Statistics, Bureau of Labor Statistics, International Labour Organization, Lux Research Inc., National Academy of Social Insurance, and the World Health Organization.

Annual Work-Related Fatalities by Industry Sector, 2007

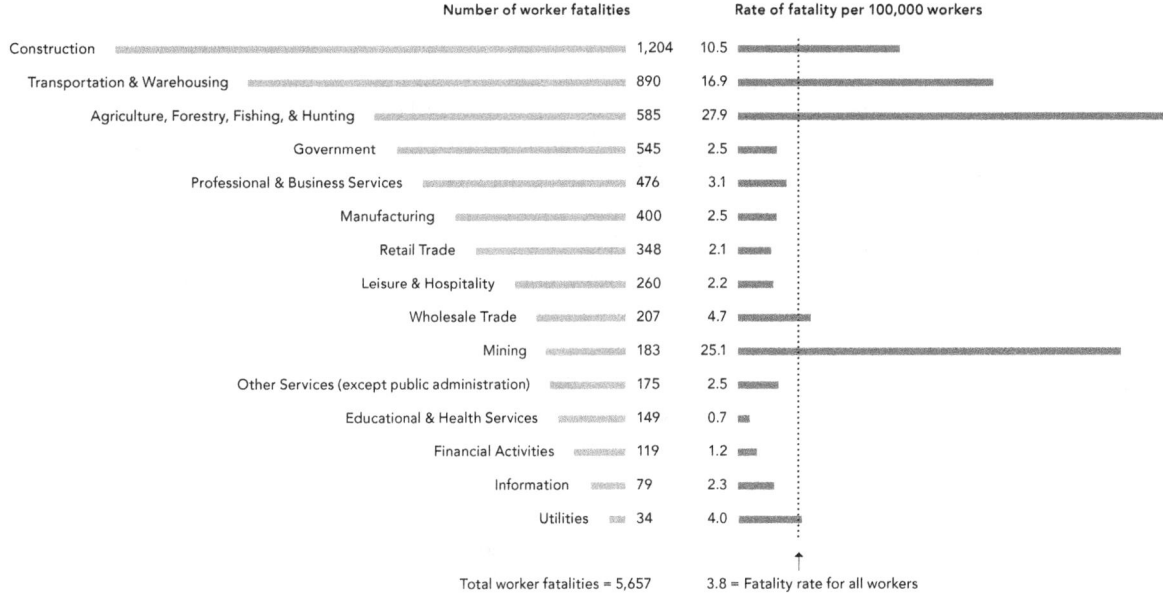

Industry	Number of worker fatalities	Rate of fatality per 100,000 workers
Construction	1,204	10.5
Transportation & Warehousing	890	16.9
Agriculture, Forestry, Fishing, & Hunting	585	27.9
Government	545	2.5
Professional & Business Services	476	3.1
Manufacturing	400	2.5
Retail Trade	348	2.1
Leisure & Hospitality	260	2.2
Wholesale Trade	207	4.7
Mining	183	25.1
Other Services (except public administration)	175	2.5
Educational & Health Services	149	0.7
Financial Activities	119	1.2
Information	79	2.3
Utilities	34	4.0

Total worker fatalities = 5,657
3.8 = Fatality rate for all workers

Annual Work-Related Fatalities by Occurrence, 2007

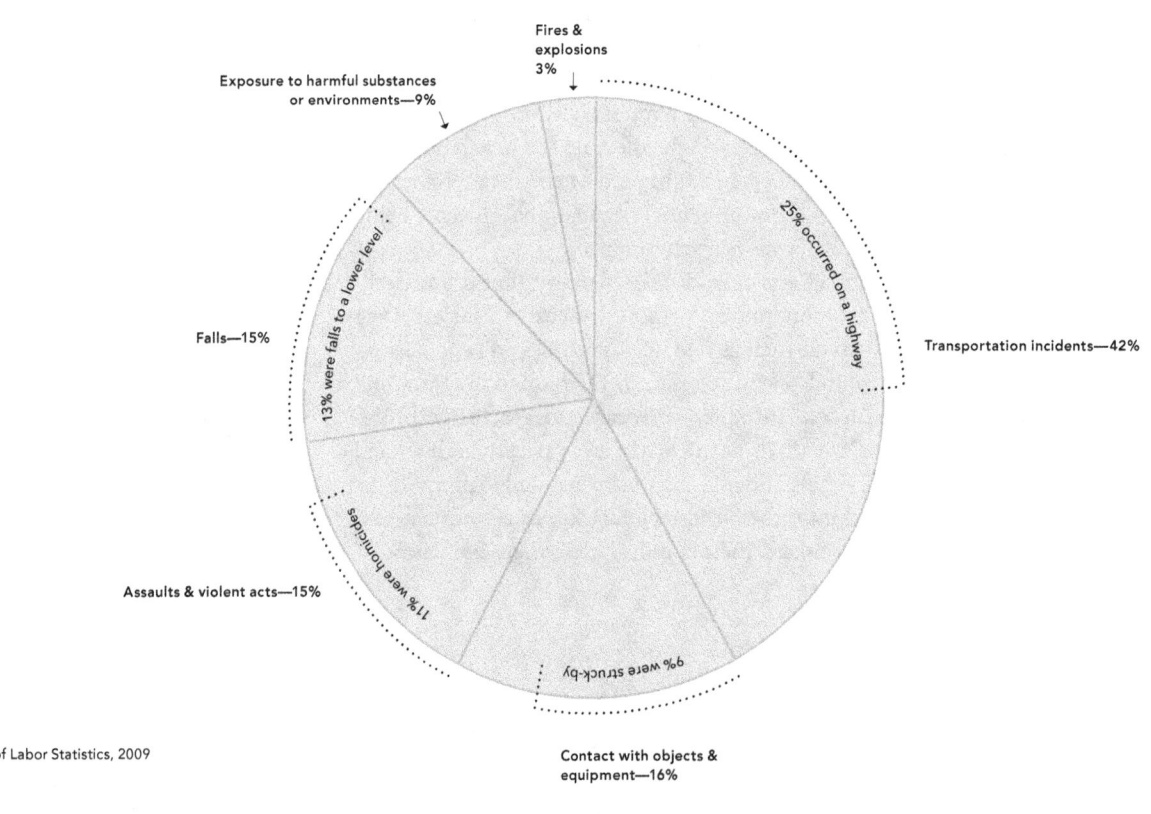

- Fires & explosions — 3%
- Exposure to harmful substances or environments — 9%
- Falls — 15% (13% were falls to a lower level)
- Assaults & violent acts — 15% (11% were homicides)
- Contact with objects & equipment — 16% (9% were struck-by)
- Transportation incidents — 42% (25% occurred on a highway)

Bureau of Labor Statistics, 2009

Days Away from Work due to Injury & Illness by Age Group, 2007

Age group	Median days away from work	Incidence rate per 10,000 workers
65 years & older	16	96
55–64 years	12	120
45–54 years	10	123
35–44 years	8	124
25–34 years	6	118
20–24 years	4	134
16–19 years	4	124

Annual Rates of Fatal Work Injuries by Age Group, 2007

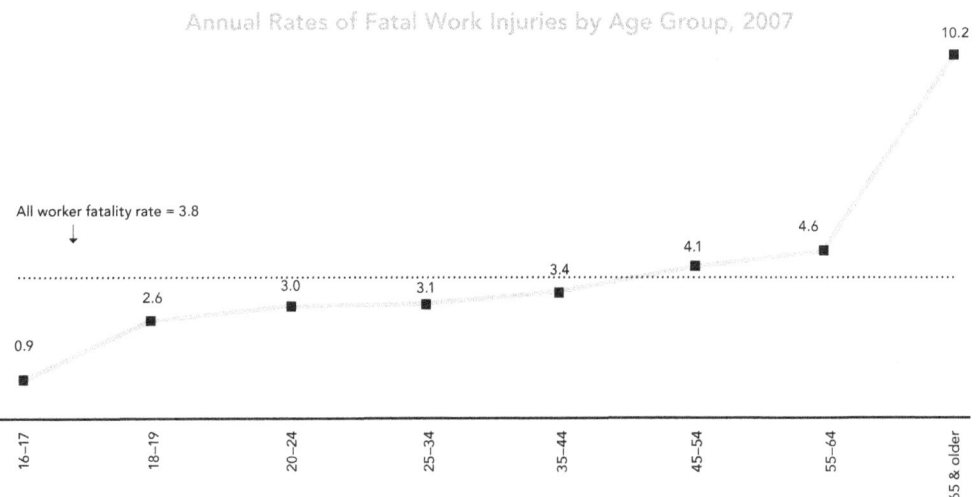

All worker fatality rate = 3.8

Age	Rate
16–17	0.9
18–19	2.6
20–24	3.0
25–34	3.1
35–44	3.4
45–54	4.1
55–64	4.6
65 & older	10.2

Annual Injury & Illness Cases Involving Hispanic or Latino Workers by Industry Sector, 2007

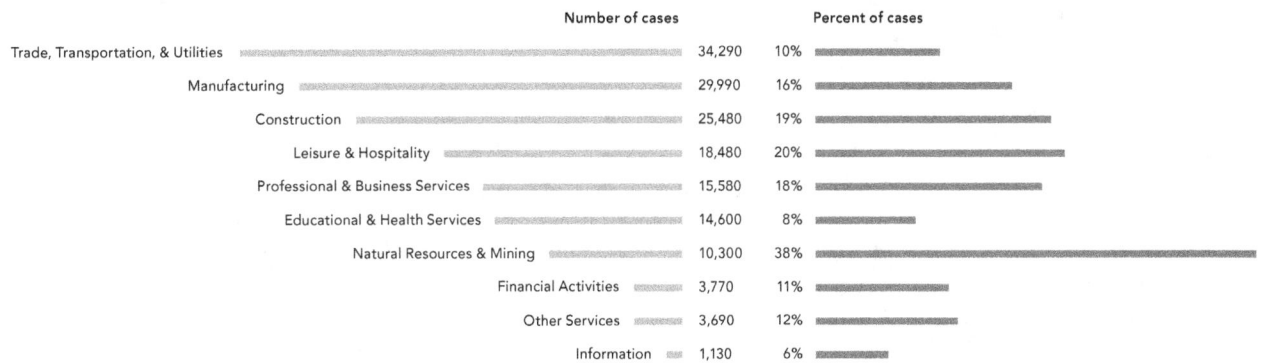

Industry Sector	Number of cases	Percent of cases
Trade, Transportation, & Utilities	34,290	10%
Manufacturing	29,990	16%
Construction	25,480	19%
Leisure & Hospitality	18,480	20%
Professional & Business Services	15,580	18%
Educational & Health Services	14,600	8%
Natural Resources & Mining	10,300	38%
Financial Activities	3,770	11%
Other Services	3,690	12%
Information	1,130	6%

Bureau of Labor Statistics, 2009

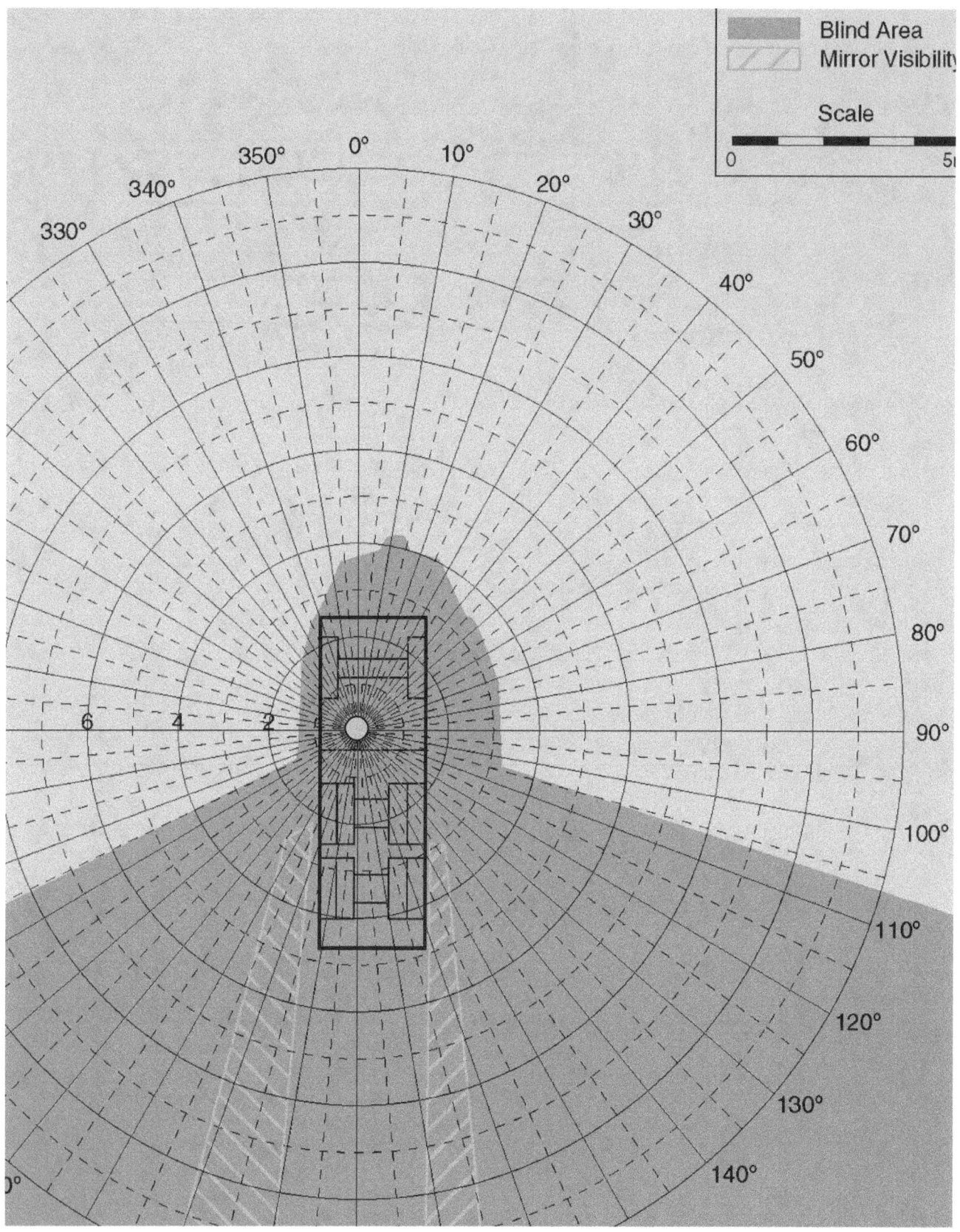

This diagram depicts the blind area of a Ford 880 construction vehicle. The truck is indicated by the black rectangle. The radial grid illustrates the 360° field of vision from the driver's seat, which is indicated by the gold dot. In the blind area, shown in grey, a driver can see nothing below the height of 4 feet 11 inches.

Illustration by David Fosbroke

Tailoring our organizational framework to fit the way people work and seek out information

National Occupational Research Agenda
Created for Stakeholders by Stakeholders

> A quick search of this document will find more than 80 instances where partnership is mentioned. Partners are a key component of NIOSH research. To date, NIOSH has entered into more than 145 formal memoranda of understanding and has engaged thousands of informal partners to assist in achieving its mission.

Workplace settings vary widely in size, design, location, work processes, workplace culture, resources, and workforce characteristics such as age, gender, training, education, cultural background, and health practices. This translates to broad diversity in the unique safety and health risks for each industry sector and the need for tailored interventions.

As a leader in occupational safety and health, NIOSH created the National Occupational Research Agenda (NORA) in 1996 as a stakeholder-driven process designed to stimulate innovative research and improved work practices. It is the framework that guides prevention efforts of the occupational safety and health community in selected priority research areas. NIOSH and its partners have been actively collaborating in this process to define research priorities for the nation and achieve impact in the prevention of work-related injury, illness, and death. NORA is a national partnership effort that has sparked unprecedented public-private accord and has created a culture of priority-driven research in occupational safety and health.

In the first decade of NORA, NIOSH and its partners developed a research agenda focused on 21 public health science priority areas that reflected considerable consensus among more than 500 groups and individuals who participated in the NORA priority-setting process.

Before entering its second decade NORA was restructured as a result of stakeholder input. NORA is now pursuing an industry sector-based approach to prioritize research needs at the national level. The North American Industry Classification System (NAICS) provides definitions for 20 sectors in order to classify businesses for the purposes of collecting, analyzing, and publishing data (visit the U.S. Census Bureau Web site for more information about NAICS). NIOSH aggregated these sectors into eight groups based on similarities in workplace safety and health issues. The eight NORA sectors are Agriculture, Forestry, and Fishing; Construction; Healthcare and Social Assistance; Manufacturing; Mining; Services; Transportation, Warehousing, and Utilities; and Wholesale and Retail Trade. NIOSH and its partners use this sector-based approach to move research results into workplace practice and to ensure a direct link between businesses, workers, researchers, and other partners to reduce work-related injury, illness, and death and improve the overall safety and health of workers.

Each NORA sector consists of a council of approximately one-third NIOSH researchers and two-thirds external partners from areas such as academia, industry, labor, and government. Each council identifies and promotes priority research needs, critical knowledge needed to fill gaps in occupational safety and health, and innovative partnerships within their sector. Currently, the councils are developing and implementing sector-based research goals, objectives, and action plans for the nation.

NIOSH manages the day-to-day operations of NORA and facilitates the work of the sector councils. In addition, NIOSH uses the knowledge, expertise, and experience of the council members to guide its internal portfolio of research programs. NORA sector programs will be assessed in light of the ongoing review of NIOSH research programs from the National Academies. For more information see pages 56 and 57.

Guiding the Implementation of the NORA Sector Approach

Strong partner involvement throughout the entire research continuum (developing a concept, planning, conducting, implementing, disseminating, and evaluating) to ensure that research is focused on achieving relevance, quality, and impact.

Access to and use of the best available surveillance data to drive research efforts, however, there is no national surveillance system in place for occupational safety and health.

A multidisciplinary approach to generate new knowledge.

The transfer and translation of research findings, technologies, and information into highly effective prevention practices and products that can be adopted immediately into the workplace.

The person symbol marks specific narrative examples of how a NIOSH or NIOSH-funded project made an impact on worker safety and health.

The clipboard symbol marks NIOSH programs that have been reviewed by the National Academies for relevance and impact. More information about the reviews can be found on pages 56–57.

Aggregating NAICS Codes into NORA Sectors

Sector	Code	Group
Agriculture, Forestry, & Fishing	11	
Construction	23	
Healthcare & Social Assistance	62	
Manufacturing	31-33	
Mining	21	
Information	51	
Finance & Insurance	52	
Real Estate, Rental, & Leasing	53	
Professional, Scientific, & Technical Services	54	
Management of Companies & Enterprises	55	
Administrative, Support, Waste Management, & Remediation	56	
Educational Services	61	Services
Arts, Entertainment, & Recreation	71	
Accommodation & Food Services	72	
Other Services (except public administration)	81	
Public Administration	92	
Transportation & Warehousing	48-49	Transportation, Warehousing, & Utilities
Utilities	22	
Retail Trade	44-45	Wholesale & Retail Trade
Wholesale Trade	42	

Agriculture, Forestry, & Fishing

The Agriculture, Forestry, and Fishing sector is the cornerstone of industries that produce and market food, fiber, and fuel. Each year it generates more than $1 trillion in economic activity, creates exports exceeding $68 billion, and provides jobs to over 2.1 million workers. In 2007 workers in this sector suffered the highest rate of fatal occupational injuries of all sectors with 27.9 deaths per 100,000 workers.

This advertisement was created with social marketing techniques that identified protecting the next generation as an important issue for farmers. It was distributed through print media in 5 New York counties. For more information visit www.cdc.gov/niosh/programs/agff/insights/insightsagrops.pdf.

Workers in the Agriculture, Forestry, and Fishing (AFF) sector perform intense seasonal work, putting in long hours in harsh physical environments, often using large machinery or being in close contact with animals. These intense working conditions lead to work-related fatality rates that are more than seven times higher for AFF workers than for workers in all industry sectors. Workers in this sector also experience a higher incidence of many adverse occupational health outcomes, including hearing loss, respiratory conditions, and skin disorders.

NIOSH and external partners have produced a formal national research agenda for occupational safety and health for the AFF sector. This agenda is based on scientific evidence, public testimonies, peer reviews, and expert consultation. It aims to highlight the most important research questions, recognize priority safety and health concerns, understand effective intervention strategies, and disseminate information on ways to achieve sustained improvements in workplace safety and health practice. The agenda outlines five strategic goals to improve worker safety and health.

Improve surveillance within the AFF sector to describe occupational hazards, worker populations at risk for adverse health outcomes, and the nature, extent, and economic burden of occupational illness, injury, and death. There is a lack of surveillance data concerning nonfatal injuries and characteristics of the workforce. Enhanced surveillance will help define specific populations at risk, injuries and illnesses of greatest concern, and the effectiveness of prevention efforts.

Reduce excessive adverse outcomes in workers who have real limits to safeguarding their own safety and health. These limits may be social factors, such as low English proficiency and literacy; socioeconomic factors; the temporary or seasonal nature of AFF work; or physiologic factors surrounding young and older workers, recent immigrant workers, and workers with physical or cognitive disabilities.

Move proven safety and health strategies into workplaces by disseminating information about relevant interventions and promoting the adoption of best practices through partnerships and collaborations.

Reduce the number, rate, and severity of traumatic injuries and deaths involving hazards of forestry, commercial fishing, and production agriculture and its support activities. In addition to more than 400 deaths, an average of 93,000 nonfatal OSHA-recordable injuries occur on U.S. farms each year. Effective measures to prevent some of these injuries are known but not widely practiced. In the forestry industry, comprehensive baseline data are needed to monitor safety conditions and track improvements. In the fishing industry, competition for a tightly controlled resource increases the likelihood that workers will take risks that could lead to injury, illness, and death. Reducing fatalities associated with vessel sinkings and falls overboard is a high priority.

Improve the health and well-being of AFF workers by reducing work-related causes or factors that contribute to acute and chronic illness and disease. Agricultural and forestry work can be repetitive, thereby leading to musculoskeletal strains and sprains. In addition, respiratory hazards, toxic chemicals, psychological stresses, and animal-borne diseases are all hazards associated with the AFF sector. New production methods and technologies, environmental issues, and changing workforce demographics continually present emerging challenges. Commercial fishing workers also face many work-related acute and chronic health exposures. More research is needed on these health issues and their prevention.

Using Marketing Research to Impact the Farming Community

One of the biggest risks to farmers is tractor overturns. Rollbars can prevent deaths from tractor overturns, however, U.S. tractors manufactured before 1985 generally do not have them. Researchers at the NIOSH-funded Northeast Center for Agricultural and Occupational Health, in partnership with the New York State legislature, implemented a rebate program and social marketing campaign (see photo on opposite page) that studied risk perceptions, barriers, and motivations to retrofitting tractors with rollbars. The campaign was effective. In the program's first 6 months, tractor dealers in 5 targeted counties sold $10^{\times}$ more rollbars than the previous 6 months and $8^{\times}$ more than dealers in counties without such a campaign.

The NIOSH Agriculture, Forestry, & Fishing Research Program has been reviewed by the National Academies (pages 56–57).

Construction

In 2008 nearly 11 million people worked at construction sites across the U.S. on any given day. Construction workers have high rates of work-related injury and death compared with other sectors. Workers in the construction sector experience the second highest rate of nonfatal injuries resulting in days away from work and the fourth highest fatality rate. They represent only about 8 percent of the U.S. workforce but experience 21 percent of the fatal injuries. Leading causes of death among construction workers are falls to a lower level; electrocutions; and struck-by, caught-in, or crushed-by incidents.

NIOSH convened a group of internal and external stakeholders from academia, industry, professional associations, labor groups, and government and nongovernment organizations to serve as council members of its NORA Construction sector research program. The mission of the Construction sector program is to eliminate occupational disease, injury, and death among workers in this industry through a focused program of research and prevention.

With guidance from available data, stakeholder input, and member expertise, the sector council identified information gaps that need to be addressed in order to protect the safety, health, and well-being of workers in the construction industry. After extensive review and discussion, the sector council developed a national agenda identifying several priority areas where focused attention is needed.

The aim of these priority areas is to reduce injury and illness from traumatic injuries, hearing loss, silica exposure, welding fume exposure, and musculoskeletal disorders; improve understanding of factors related to work organization, organizational culture, safety and health management systems, and occupational health disparities; develop, strengthen, and expand training and education; increase the use of prevention through design strategies; and improve surveillance at the federal, state, and private levels.

The national agenda provides a framework for sector stakeholders to share what they believe are the most relevant issues, gaps, and safety and health needs in the construction industry and provides guidance to prioritize their work among the many competing safety and health concerns. It is intended to inspire decision makers to adopt these issues as their top priorities, steer researchers to cohesive topic areas for research proposals, and encourage dialog and partnering among stakeholders on a subset of key issues—thus ensuring an increased focus on reducing injury, illness, and death among construction workers.

Stakeholder audiences that may be particularly interested in the Construction sector national agenda include research funding sources, such as federal and state research agencies, building-owner associations, and research organizations for workers' compensation insurance companies; public and private researchers; construction industry organizations, such as tool, equipment, and material manufacturers and distributors, trade associations, apprenticeship training organizations, and management firms; and safety and health practitioners.

The NORA Construction sector council encourages stakeholders in the construction industry to collaborate with NIOSH and the council to address the strategic goals outlined in its national agenda. Several opportunities are available for collaboration, including identifying, developing, and implementing safety-related devices, methods, or systems through partnerships with organizations and companies that will actively support these solutions during normal construction activities. A well-designed and facilitated approach to address these important safety and health issues can increase the likelihood of reducing risks faced daily by construction workers.

A construction worker leans against the guardrail during a field trial of the Adjustable Guardrail Bracket. The unit is attached to a sloped residential roof.
Photo by Thomas G. Bobick

Impacting Worker Safety Through Intervention & Product Design

Current fall protection products in the construction industry are limited to flat surfaces, are adjustable only for a few roof pitches, or are lacking components to support a horizontal guardrail. NIOSH engineers designed an adjustable roof bracket-safety rail assembly that can prevent construction workers from falling from roof edges or through roof openings and skylights. The adjustable design allows for installation of a protective guardrail system on flat commercial and industrial roofs and on residential roofs with 7 different slopes ranging from 27°–63°. Two OSHA regulations were the basis for the initial design. The prototype was improved using laboratory performance testing and user feedback and is patent-pending in the U.S. and Canada. Once commercially available, this fall-prevention technology will have immediate potential to reduce the frequency of traumatic injuries experienced by those working at heights. Approximately 37% (447) of the work-related deaths among construction workers resulted from a fall in 2007.

Relevant Information

In 2007, > 1,200 construction workers died from a work-related injury and > 380,000 nonfatal injuries and illnesses were recorded.

Falls, electrical hazards, and contact with objects and equipment account for about 63% of fatalities among construction workers.

In 2007, approximately 200,000 cases of nonfatal injury and illness were reported among construction workers that resulted in days away from work, job transfer, or restriction. 42% of these workers were laborers and carpenters.

The NIOSH Construction Research Program has been reviewed by the National Academies (pages 56–57).

Healthcare & Social Assistance

Many people think of the Healthcare and Social Assistance industries as being clean, sterile, and safe places to work. In reality, workers in this sector are exposed to many hazards that can adversely affect their health and well-being. Long hours, changing shifts, physically demanding tasks, violence, and exposures to infectious diseases and harmful chemicals are examples of hazards that put these workers at risk for illness and injury. More than 18 million workers in the Healthcare and Social Assistance sector face hazards as they work to serve the sick and those in need.

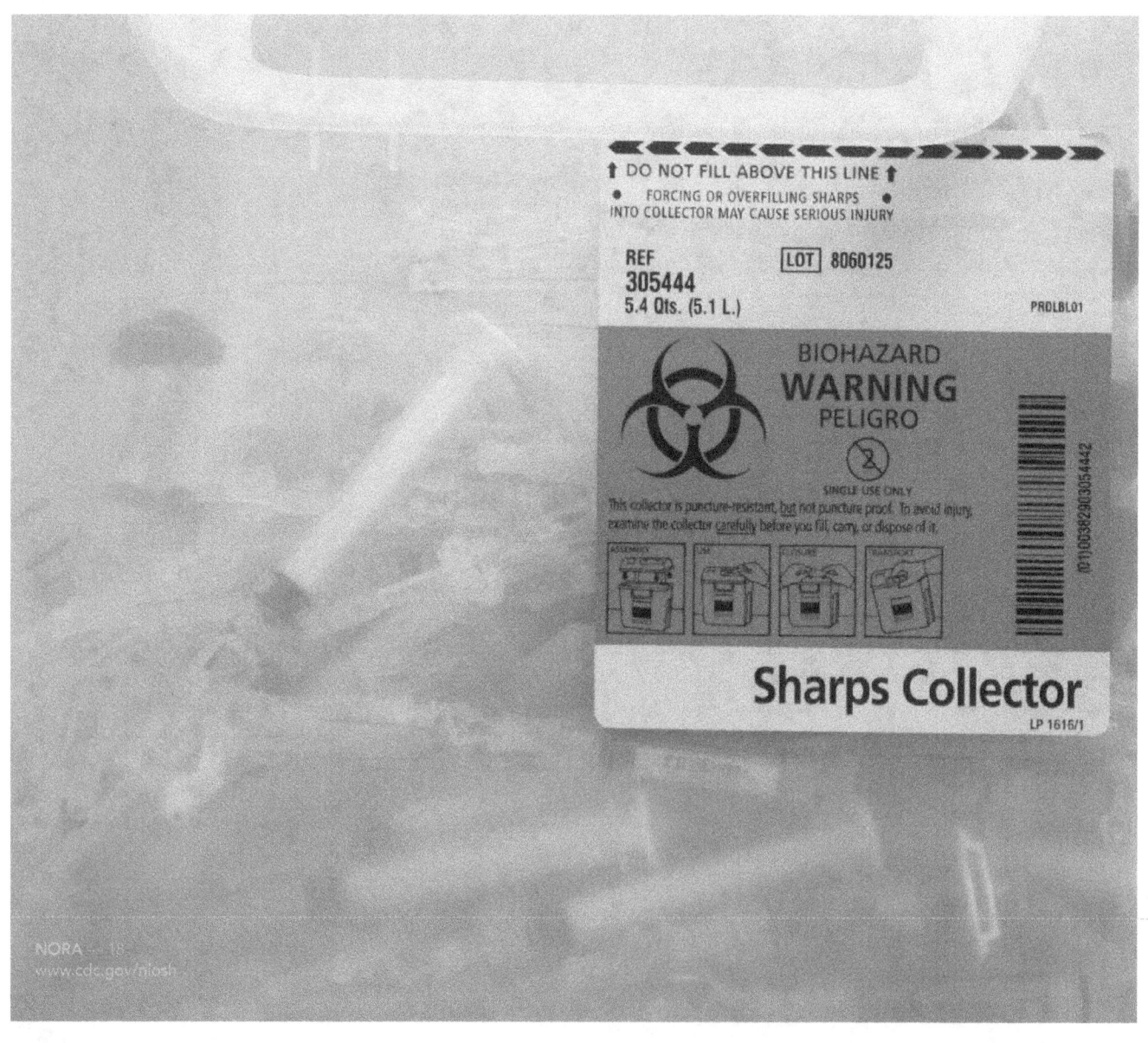

The Healthcare and Social Assistance (HSA) sector, as defined by NORA, includes workplace establishments that provide healthcare and social assistance for individuals. Because it can be difficult at times to clearly define the boundaries between healthcare and social assistance, the sector combines them. The industries in this sector are arranged on a continuum starting with those that provide medical care exclusively, to those that provide both healthcare and social assistance, to those that provide social assistance only.

The mission of the NORA HSA sector is to eliminate occupational disease, injury, and death among individuals working in this sector through a focused program of research and prevention. To accomplish this mission, NIOSH developed a council of experts from inside and outside its Institute. The council developed six strategic goals as part of its agenda for the nation.

Conduct surveillance to better understand exposure to hazards, injuries, illnesses, and "near-misses." This information should routinely be collected and analyzed because it is very important for setting priorities and making improvements to protect workers within this sector.

Promote the use of existing approaches, techniques, and safe work practices, such as lifting equipment to aid with patient transfers, safer medical devices, and reasonable work schedules and workloads, to protect HSA workers from injury and illness. These techniques can protect workers from back and shoulder injuries, reduce the likelihood of suffering a sharps injury, or reduce stress and medical errors. Safe work practices can be promoted through demonstration projects, intervention effectiveness studies, and public health marketing. These approaches aid in learning and refining best practices for specific work settings, show financial benefits and patient safety improvements, and raise awareness of ways to address occupational safety and health issues.

Promote safety and health management programs to reduce injury and illness rates. Comprehensive safety and health programs can also realize a return-on-investment of more than $3 for every $1 invested. All HSA sector employers should have such a program in place.

Conduct research and develop safe approaches to protect HSA workers. Improvements, such as eliminating hazardous materials, using safer equipment and work practices, and taking advantage of new vaccines and preventive treatments for infectious exposures, are needed to ensure the safety and health of workers in this sector.

Address all healthcare settings to ensure workers employed in nonhospital settings are also protected from injury and illness. About half of all healthcare workers are employed in nonhospital settings, half of which are small businesses with fewer than 20 employees. In addition, care is often provided in homes. These environments are frequently more uncontrolled and can lead to worker injury, illness, or death.

Build strong partnerships among industry, labor, academia, government, and those who have a stake in protecting the safety, health, and well-being of patients and workers in the HSA sector. The sector council encourages interested stakeholders to collaborate with NIOSH and the council to advance the mission of the NORA HSA sector program.

Research & Recommendations Impact the Healthcare Industry

Because of the HIV epidemic, healthcare workers used far greater numbers of powdered latex gloves in the 1990s than in previous years. This increased use resulted in many more workers exposed to powders that carried latex allergens. Healthcare workers presented with latex allergies in epidemic proportions: hives, work-related asthma, and even anaphylaxis. In an Alert NIOSH advised, among other recommendations, that healthcare establishments and healthcare workers reduce the use of powdered latex gloves to minimize latex allergen exposure. NIOSH evaluated the effectiveness of the Alert and found that hospitals that received the Alert were more likely to have progressed from a position of "inaction" to "advocacy" on the issue of latex and latex use. Directors who had seen the Alert were more likely to report that they intended to implement an employee health education program, replace powdered latex gloves, and restrict latex glove use as compared with directors who had not seen it. NIOSH was one of the first organizations to recommend avoiding powdered latex gloves. Eventually many other organizations called for this change in the healthcare industry. Research has shown a shift away from the use of powdered latex gloves to nonpowdered latex or nonlatex gloves. As a result, allergic reactions to latex among healthcare workers have been markedly reduced.

Manufacturing

The Manufacturing sector, made up of 21 industries, has a national focus on becoming more competitive at home and abroad. Of the 8 NORA sectors, manufacturing faces the most diversity in workplace safety and health challenges. Hearing loss, nanotechnology, chemical process safety, management systems, and special populations are some of the priority research needs to protect the nearly 16 million U.S. workers in this sector.

Many corporations recognize the significance of including workplace safety and health in their corporate strategy for producing better products, improving the global economy, and maintaining a healthy workforce.

Women make up 29 percent of workers in the Manufacturing sector. About 31 percent are ethnic minorities (15 percent Hispanic or Latino, 10 percent Black or African American, and 6 percent Asian). Occupations include production workers, inspectors, machinists, purchasing agents, and assemblers.

In 2007 work-related injuries claimed the lives of 400 workers in the Manufacturing sector. The leading causes of death were contact with objects and equipment, transportation incidents, and falls. About 783,000 nonfatal injuries and illnesses were reported. Of these, more than 187,000 resulted in days away from work and an additional 239,900 required a job transfer or restriction.

The leading causes of days away from work were contact with objects or equipment, overexertion, repetitive motion, and falls. Of all industries, 14 reported more than 100,000 nonfatal injuries and illnesses; 3 of these were among the largest of the manufacturing industries—transportation equipment manufacturing, fabricated metal product manufacturing, and food manufacturing.

To address these and other worker safety and health concerns, NIOSH developed a coordinated research program in 2006 as part of its NORA framework. The mission of the NORA Manufacturing sector program is to eliminate occupational disease, injury, and death among people working in the Manufacturing sector. NIOSH established a council of external and internal experts to identify the top occupational safety and health problems, research gaps, emerging issues, and relevant priority populations.

The sector council is made up of stakeholders from industry, government, labor, and academia. With guidance from available data; council member expertise; and input from workers, employers, professional associations, and other stakeholders, the sector council is developing a research agenda for the nation that identifies priority goals to protect the safety, health, and well-being of workers in the manufacturing sector.

Priority areas of research and translation identified by the Manufacturing sector council include contact with equipment and objects; slips, trips, and falls; ergonomics that can lead to musculoskeletal disorders; hearing loss prevention; chemical exposure; work organizational systems and stress; special populations such as contract workers, younger and older workers, immigrants, and women of child-bearing age; nanotechnology; and small business.

The Manufacturing sector council encourages stakeholders in the manufacturing industry to collaborate with NIOSH and the council to address the strategic goals outlined in its national agenda. Partners are essential to conducting new research, assessing the state of the field, and communicating findings to those in positions who can make positive changes in the workplace and reduce injury, illness, and death.

North American Industry Classification System
(codes 31-33)
311 -- Food Manufacturing
312 -- Beverage & Tobacco Product Manufacturing
313 -- Textile Mills
314 -- Textile Product Mills
315 -- Apparel Manufacturing
316 -- Leather & Allied Product Manufacturing
321 -- Wood Product Manufacturing
322 -- Paper Manufacturing
323 -- Printing & Related Support Activities
324 -- Petroleum & Coal Products Manufacturing
325 -- Chemical Manufacturing
326 -- Plastics & Rubber Products Manufacturing
327 -- Nonmetallic Mineral Product Manufacturing
331 -- Primary Metal Manufacturing
332 -- Fabricated Metal Product Manufacturing
333 -- Machinery Manufacturing
334 -- Computer & Electronic Product Manufacturing
335 -- Electrical Equipment, Appliance,
       & Component Manufacturing
336 -- Transportation Equipment Manufacturing
337 -- Furniture & Related Product Manufacturing
339 -- Miscellaneous Manufacturing

Identifying Emerging Hazards & Impacting Worker Health Through State-Based Partnerships

Bronchiolitis obliterans, a rare and life-threatening lung disease, was found among microwave popcorn manufacturing workers and linked by NIOSH to the chemical diacetyl, an ingredient in artificial butter flavorings. At the annual meeting of the Council of State and Territorial Epidemiologists in June 2006, the California state health department and NIOSH officials reported a similar outbreak of lung disease among 3 food flavorings manufacturing workers exposed to diacetyl. To respond to this emerging hazard, members of NIOSH's 15-state Consortium of Occupational State-based Surveillance group conducted an array of occupational public health activities. State health officials have informed workers, employers, and healthcare providers of the risks from work-related exposure to food flavorings; have conducted worksite visits or employer surveys to assess diacetyl and the use of food flavorings; have helped state OSHA programs identify worksites that are using food flavorings; and have responded to reports of possible health risks from use of diacetyl-containing oils in restaurants.

Mining

Despite the significant progress made to protect the safety and health of miners, to date the mining industry's fatality rate remains more than 6 times the national average for all occupational sectors. High rates of lost work time and permanently disabling injuries plague the industry as do diseases from long-term exposure, such as respiratory illness and hearing impairment.

Mining, as a basic industry, provides the foundation for our nation's economy, national defense, and standard of living. The mining industry spans all 50 states and our country depends heavily upon it for raw materials and energy. Wire, pipe, concrete, steel, and glass serve as examples of society's reliance on minerals. Many of us, however, fail to realize that hundreds of mined commodities play integral roles in everyday products, from medicines to computers. Furthermore, coal supplies a large portion of our nation's energy requirement, accounting for more than 50 percent of all generated electric power. Domestic mining also provides strategic minerals that are important for our national security.

Extracting minerals from the earth involves complex, unpredictable, and often labor-intensive processes. As a result, the mining industry historically experiences high risks of injury and death. The most challenging working conditions occur in the industry's underground sector and often include confined spaces, interactions with large mobile equipment, poor lighting, wet surroundings, explosive gas, airborne dust, high noise levels, and uncertain roof conditions. Despite significant advancements through the decades, there remains room for improvement as mining conditions become more complex. Adding to the challenge is the fact that a generation of seasoned miners is retiring.

The recent disasters at the Sago, Alma, and Crandall Canyon Mines further exemplify the safety hazards and challenges associated with mining. The MINER Act of 2006 created the Office of Mine Safety and Health Research within NIOSH, providing a focused structure for the Institute's research in the Mining sector.

The mission of the NORA Mining sector research program is to eliminate work-related disease, injury, and death from the mining workplace. In collaboration with external partners, NIOSH developed a set of priority research goals to achieve this mission. Priorities were established using surveillance data, stakeholder input, loss control, and the Mining Safety and Health Research Advisory Council.

Priority research goals are aimed at reducing health hazards in the workplace to reduce respiratory disease; noise-induced hearing loss; repetitive and cumulative musculoskeletal injuries; traumatic injuries; the risk of mine disasters, such as fires, explosions, and inundations; and ground failure injury and death. Minimizing the risk and enhancing the safety and effectiveness of emergency responders is also of high importance. It is essential to determine the impact of changing mining conditions, new and emerging technologies, and the shifting patterns of work on worker safety and health to prevent potentially adverse outcomes.

One significant change in the Mining sector is the growth in production and the number of operations resulting from the rapidly increasing demand for affordable energy. For example, the oil and gas subsector, in 2003—at the beginning of the current boom cycle—employed 292,846 workers. By 2007 employment had grown more than 46 percent and the annual number of drilling and workover rigs increased by 39 percent. Unfortunately the number of deaths in that same period also increased from 85 to 122. Despite this significant increase, the fatality rate remained steady at 30 deaths per 100,000 workers. This is almost 8 times the rate for all U.S. workers and represents about two-thirds of all mining fatalities. This is a high-risk industry; if the demand for oil and gas remains high the death rate will also remain high unless effective safety programs are implemented. NIOSH is working collaboratively with industry and others to identify and solve safety problems through research and prevention activities.

Impacting the Mining Community Through Extensive Research & Intervention Efforts

The National Academies' review of the NIOSH Mining Program underscores its significant contributions to improving mine worker safety and health. The research, technology, and training products developed by NIOSH have been widely adopted and used in mines throughout the country. This program has a long history of meeting the safety and health needs of its customers and stakeholders and continues to receive strong support from its labor and industry constituents. Notable examples of the range of NIOSH's impact include computer models that underground mining companies use to safely design mine openings; a training program for safely donning self-contained self rescuers; newly developed engineering controls for mining machines that are being adopted by manufacturers for reducing noise levels; scientific studies that have formed the basis for new safety and health regulations in critical areas such as mine seals and refuge chambers; design guides and tools that are widely applied in underground mines for reducing harmful diesel emissions; real-time monitoring instruments; and the recently developed postaccident communication technologies that are enabling the underground coal industry to comply with requirements of the MINER Act of 2006.

The NIOSH Mining Research Program has been reviewed by the National Academies (pages 56–57).

Relevant Information

More than 75% of mine workers suffer from occupational hearing loss by the time they retire; many of these workers also suffer from chronic musculoskeletal disorders.
The average age of mine workers is approaching 50 years.
An influx of inexperienced miners creates its own set of challenges and opportunities with respect to safety.

Photo by Elaine Cullen

Services

In 2008 the NORA Services sector employed nearly 69 million workers. This sector is the most diverse in its job categories and includes industries such as public safety and other government services, automotive repair, hotels and restaurants, education, recreation, and waste collection. U.S. workers in the Services sector are exposed to a variety of substances such as asbestos, cleaning solvents, carbon monoxide, diesel exhaust, and tobacco smoke; physical stressors such as temperature extremes, overexertion, assaults, and other forms of violence; and safety hazards such as electrocution, motor vehicle crashes, and slips, trips, and falls.

Work environments vary widely in the Services sector and include offices, hotel rooms, indoor and outdoor entertainment facilities, restaurant kitchens, classrooms, automotive garages, public roads, and private households. These varied—and often uncontrolled—environments put workers at increased risk of workplace injury and illness.

From 2003–2007 homicides and other violent acts were the leading causes of death among workers in food and alcohol service, hotel and motel, and residential leasing. In 2006 lost workday injury rates were greatest for landscapers, waste collectors, professional athletes, and those employed in consumer electronics and appliance rental; musculoskeletal disorders from workplace activity were elevated for workers in residential real estate leasing, building services, landscape, and laundry services; and fatalities among public safety workers from vehicle crashes, homicides, and heart disease had risen.

To address occupational safety and health issues in the Services sector, NIOSH established a NORA sector council in 2006 that is comprised of industry experts, labor representatives, academic investigators, and public health practitioners. Because public safety workers experience many unique occupational hazards, a separate council was convened with experts from law enforcement, fire fighting, corrections, and emergency medical services. Each council has identified gaps in research, knowledge, and interventions that were used to develop a national agenda. This agenda includes priority research and prevention goals for each of the 14 industries encompassed within the NORA Services sector.

Priority goals address important safety and health issues, such as roadway hazards and vehicle crashes for outdoor workers; workplace violence and stress for law enforcement, corrections, and restaurant workers; overexertion and falls from heights for solid waste and landscape workers; effective training programs for youth and immigrant workers and workers with disabilities; fire ground exposures and chronic disease among fire fighters; and patient transfer equipment for emergency medical service workers.

These goals and other Service sector activities will be accomplished through internal and external—and new and existing—partnerships. Through the cooperative efforts of workers, management, labor, practitioners, and scientists, NIOSH will continue to reduce the economic and personal impact of occupational disease, disability, and death through high-quality research and effective prevention strategies, as well as help ensure the success of research translation, reduce work-related hazards and exposure, and protect workers in the Services sector.

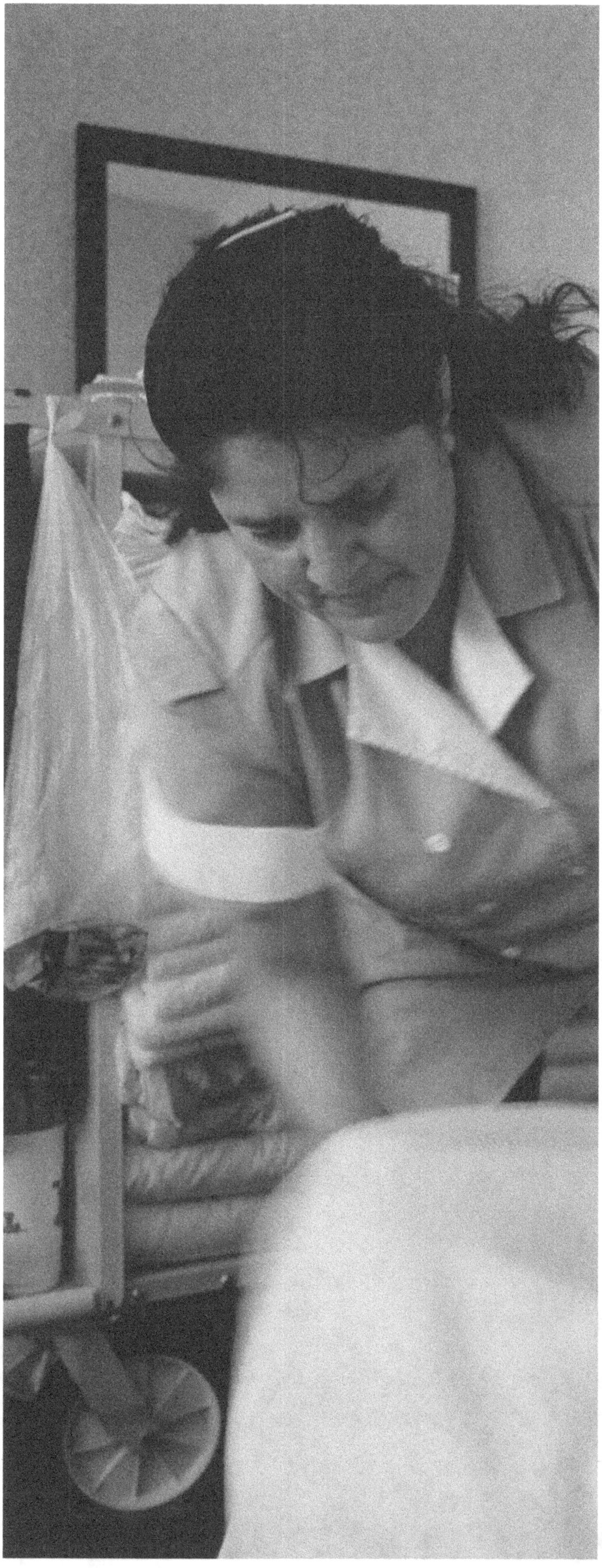

Protecting & Impacting Those Who Protect Us

NIOSH occupational research and prevention products have impacted many of the industries within the NORA Services sector. For example, Health Hazard Evaluations have resulted in a manual developed by the U.S. Fire Administration on hearing loss prevention during response and training exercises, in a Vermont Public Service Board agreement to eliminate worker exposure to creosote through substituting creosote-treated electrical poles with pressure-treated ones, and in a training course required for Canadian Boat Licensure. In addition, the Fire Fighter Fatality Investigation and Prevention Program at NIOSH has contributed to the development of a National Fire Protection Association Standard on fire fighter staffing levels and an investigation cited as justification for passing the Bradley Law, which changed fire fighter training procedures in New York State.

See page 40 for more information about Health Hazard Evaluations and page 61 for the Fire Fighter Fatality Investigation and Prevention Program.

Relevant Information

In 2007, > 1.5 million nonfatal injuries and illnesses were reported among Service sector workers; 18% of these cases resulted in days away from work.
The number of fatal workplace injuries among protective service workers rose 19% to 337. This was led by an increase in the number of police officers fatally injured on the job.
In 2007, homicides (628) were the 3rd leading cause of work-related fatalities and outnumbered struck-by fatalities (504) for the first time since 2003.
Throughout the next decade, U.S. employment growth in education (32%), professional and business services (28%), and leisure and hospitality (18%) are projected to be well-above the national projection of 15%.

Transportation, Warehousing, & Utilities

In 2007 the NORA Transportation, Warehousing, and Utilities sector accounted for 16 percent of all workplace deaths. Transportation incidents accounted for 71 percent of the fatalities in this sector. The subsectors of air transportation, couriers, truck transportation, and warehousing had rates of lost-workday injuries from overexertion that were 2–8 times higher than those for all U.S. workers. Traumatic injuries, musculoskeletal disorders, and hazardous exposures are the primary concerns for affecting the safety and health of workers in this sector.

The Transportation, Warehousing, and Utilities (TWU) sector is very diverse in its workforce, job responsibilities, and hazards. The sector covers all modes of transporting passengers and cargo—air, rail, water, road, and pipeline—as well as support activities related to all modes; establishments engaged primarily in warehousing and storage of goods; and electric power, natural gas, water, sewage, and other systems.

NIOSH convened a NORA sector council to assist in identifying the highest priority research areas within the TWU sector to improve the safety and health of its workers. The mission of the TWU sector program is to implement a focused program of surveillance, research, and prevention that leads to a sustained reduction in work-related injury, illness, and death among the TWU workforce. In striving to achieve this mission, the council created a research agenda that delineates national priorities for the TWU sector. The agenda identifies research, information, and actions most urgently needed to further prevention of work-related injury and illness, and it provides a vehicle for stakeholders to describe the sector's most relevant issues, gaps, and safety and health needs.

Four research areas were identified where focused efforts are needed: traumatic injury and death rates that result in lost workdays; incidence and severity of work-related musculoskeletal disorders; workplace programs and practices that enable workers to engage in healthy behaviors to reduce work-related physiological and psychological stressors, improve the use of healthcare services, and reduce premature death; and chemical, biological, physical, and psychosocial work-related hazards and exposures.

Within each priority area, the TWU sector council identified goals and projects to address safety and health issues more specifically within its many subsectors. Examples involving workers in air transportation, ground transportation, storage operations, and utilities include investigating the relationship between breast cancer and female flight attendants, worker exposure to jet fuel, and the relationship between airflow in commercial aircraft cabins and potential disease-causing particles; investigating truck driver fatigue and mortality of independent owner-operator truck drivers, and developing methods to allow taxi drivers to warn others that they are in danger of assault; incorporating effective interventions into industry policies and procedures to prevent injuries from contact with objects and equipment or slips, trips, and falls on working surfaces, and establishing effective interventions and best practices to prevent work-related musculoskeletal disorders; and identifying risk factors for electrical and chemical burns from working with high or low-voltage panel apparatus and conductors, cut, crush, and amputation injuries from energized equipment and tools, and injuries from falls to a lower level while working on poles, towers, platforms, ladders, and scaffolding.

The TWU sector agenda is intended to assist stakeholders, such as industry, labor, safety professionals, and academics, in prioritizing their work among the many safety and health issues of interest; inspire decision makers and program planners to include these topics in their top priorities; guide researchers to relevant topic areas for research proposals; and encourage dialogue and partnering among stakeholders on a subset of key issues—thereby increasing our collective ability to effectively reduce injuries and illnesses among workers in the TWU sector.

Impacting the Safety of Transportation Workers in the U.S. and Globally

The TWU research program is young in comparison with many other NIOSH programs. Consequently, many of the TWU projects are focused on identifying and understanding injuries and illnesses to support the development of prevention efforts. Efforts have been made, however, to positively impact and reduce injuries and illnesses in this sector. For example, the TWU program was represented during the development of the American National Standards Institute standard Z15.1, *Safe Practices for Motor Vehicle Operations*, which delineates minimum requirements for workplace traffic safety programs. From an international perspective, TWU representation provided input into a United Nations General Assembly Resolution on "Improving Global Road Safety." This resolution was adopted on March 31, 2008 with full U.S. support.

Relevant Information

In 2007, > 7.7 million people worked in the TWU sector. Of these workers 924 fatalities from work-related injury or illness, and more than 289,500 nonfatal injuries and illnesses were reported among TWU workers.

Approximately 65% of the nonfatal injuries and illnesses reported among workers in the TWU sector in 2007 resulted in days away from work, job transfer, or restriction.

The TWU program is closely linked with the NIOSH Global Collaborations Program, which focuses on reducing work-related disease, injury, and death among all workers employed globally.

Wholesale & Retail Trade

The Wholesale and Retail Trade sector, as defined by NORA, employs approximately 20.5 million workers and is expected to grow by 12 percent throughout the next decade. In 2007 555 workers died in the Wholesale and Retail Trade sector and more than 822,000 cases of nonfatal injury and illness were reported. Approximately 54 percent of these cases required days away from work, job transfer, or restriction. The sector's cumulative burden of occupational injury, illness, and death is among the largest in the U.S.

The Wholesale and Retail Trade (WRT) sector is comprised of approximately 1.6 million businesses that range from one-person, one-location worksites to a 1.7 million-employee chain store with more than 3,000 worksites. Changes in the business climate, consumer demands, and new technologies lead to frequent changes in job conditions. WRT workers are increasingly contract, temporary, or part-time. More than 70 percent of WRT workers are employed with small businesses. The sector typically employs the youngest and oldest workers, many of whom are culturally and linguistically diverse. Jobs in WRT often have significant turnover, require little or no education, are entry-level, and offer low wages. In addition, people perceive jobs in this sector to be safe, which undermines the need for safety training. These and other characteristics of the WRT sector can negatively affect the safety and health of workers and put them at increased risk of workplace injury and illness.

The mission of the NORA WRT sector program is to promote a safe and healthy workplace for all WRT workers by preventing work-related injury and illness through research and public health practices. In 2006 NIOSH convened a WRT sector council comprised of industry experts, labor representatives, academic investigators, and public health practitioners. After the council's review of available occupational safety and health surveillance data and input from stakeholders, it identified existing gaps in knowledge and interventions and addressed the question "what information is needed to more effectively prevent injury and illness in the WRT sector?" As a result of these efforts, the council developed six strategic goals designed to address the sector's primary safety and health concerns.

The strategic goals aim to reduce chronic musculoskeletal disorders, traumatic injuries caused by slips, trips, and falls, acute injuries from contact with hard objects, workplace violence, and motor vehicle-related injury and death; improve outreach to small businesses; increase understanding of how vulnerable populations experience disproportionate risks; and expand the availability and use of effective interventions to reduce injury and illness among this population.

A comprehensive and focused effort involving surveillance, needs assessment, data management, economic analyses, research, interventions, information dissemination, and evaluation is needed to accomplish the goals identified by the WRT sector council. These goals are intended to be accomplished through new and existing partnerships between NIOSH and its stakeholders over the course of the next decade.

Within the WRT sector, workers are potentially exposed to an assortment of occupational hazards including psychosocial factors (i.e., stress from lack of job security and frequent interaction with the public), long workdays, shift work, violence, materials handling, static posture, prolonged standing, repetitive motion, and heavy lifting. Although the potential hazards are varied and the incidence rates are high, BLS suggests that the overall number of injuries and fatalities within the WRT sector may be attributed to a specific subset of high-risk workplaces such as mail order, home stores, and gas stations.

Impacting a Workforce Exposed to a Variety of Diverse Hazards

NIOSH occupational research and prevention activities have led to several successful outcomes in the WRT sector. Examples include violence prevention programs, which provided training and information to small businesses to reduce injury and illness from workplace violence and reduced fatalities by 60% among gas station workers; research on slips, trips, and falls in warehouse distribution centers, which reduced injuries among order pickers in grocery warehouses by 40%; and ventilation studies, which led to the installation of improved dust exhaust hoods in retail lumber stores and reduced reports of lung disorders by 40% among workers.

Relevant Information

About 90% of Wholesale establishments employ < 20 workers.

Workers from 16–19 years of age have 2× the injury rate of all workers. Many in this age group work in the WRT sector.

From 1992–2001, 46% (3,637) of all homicides occurred in Retail Trade. The cost to society was estimated at $2.6 billion.

 ON PARTNERSHIP

Enhancing & Advancing NIOSH Research Through Partnerships

NIOSH is dedicated to developing and maintaining long-lasting and productive partnerships with people and organizations committed to protecting the safety, health, and well-being of workers and with those whose efforts help to promote the movement of NIOSH research into practice to prevent work-related injury, illness, and death. Partnerships are essential in demonstrating and sustaining positive impact.

"What does it mean to you to partner with NIOSH?"

We asked 12 of our partners to share their thoughts on this question. Their candid responses are provided on the following pages.

Partners are pictured from left to right, then top to bottom

Melissa McDiarmid, Professor—University of Maryland School of Medicine | **Bruce Watzman**, Senior Vice President for Regulatory Affairs—National Mining Association | **Mei-Li Lin**, Senior Director—National Safety Council and Editor—Journal of Safety Research | **Joseph Fowler, Jr.**, Executive Director—Laborers' Health and Safety Fund of North America | **Nicholas Mastrodicasa**, Project Manager for the State of Alaska Department of Transportation—Aviation Division | **Peter O'Neil**, Executive Director—American Industrial Hygiene Association | **Tom Leamon**, Adjunct Professor of Occupational Safety—Harvard School of Public Health | **Ginny Frings**, Visiting Professor—Xavier University and President—Concerned Angels, Inc. | **James Koskan**, Corporate Director Risk Control—SUPERVALU INC | **Dave Heidorn**, Manager for Government Affairs and Policy—American Society of Safety Engineers | **Frank Renshaw**, Senior Environmental Health and Safety Leader—The Dow Chemical Company | **Barbara Lee**, Director—National Children's Center for Rural and Agricultural Health and Safety

Barbara Lee—In 1992 the Director of NIOSH participated in the first national symposium addressing childhood agricultural injury prevention. Since then, every NIOSH leader has supported interventions that address children on farms regardless of their work or residency status. Data now confirm that the rate of childhood agricultural injuries has dropped by nearly 50 percent. NIOSH leadership, combined with private-sector commitment and influence, should be credited for this success. It is personally and professionally rewarding to be a part of this movement impacting the well-being of children.

Dave Heidorn—The relationships that have been developed between NIOSH, ASSE, and its members through our partnership have fostered an increased sense of shared responsibility for the role that safety plays in NIOSH research and education. Our members are now far more engaged in the NORA research agenda activities. They have participated on the NIOSH Board of Scientific Counselors, have been involved in determining the direction of training grants, and have benefitted from increased involvement of NIOSH staff in ASSE educational programs. Members have also benefitted from NIOSH staff publishing in ASSE communications. In return, ASSE's members have developed a greater understanding and increased support for NIOSH's valuable role in preventing work-related injury, illness, and death. I believe NIOSH has also benefitted through increased exposure to our members' experience in facing safety and health risks on the job, particularly in areas such as ergonomics and design. ASSE has also provided NIOSH staff the opportunity to be a part of the safety and health professional community where they belong. We look forward to future opportunities to build on this partnership.

Joseph Fowler, Jr.—For the past 21 years, the Laborers' Health and Safety Fund of North America has had an active and very productive partnership with NIOSH. The Fund has collaborated with NIOSH to reduce fumes from asphalt pavers, prevent hearing loss among construction workers, prevent run-overs and back-overs of workers by construction vehicles, and respond to toxic exposures. The Fund also works actively with NIOSH on medical surveillance efforts for workers who responded to the World Trade Center disaster, dose reconstructions to help energy workers receive the compensation they deserve, and research activities for the NORA Construction sector program. These efforts are just beginning to pay off as more research will focus on broad issues, such as how the structure of the industry (i.e., low-bid contracts) affects the safety and health of workers and how interventions can be designed for maximum impact.

Melissa McDiarmid—Almost 25 years ago I received scholarship support from NIOSH for my Master of Public Health degree and Occupational Medicine training through the Johns Hopkins Education and Research Center. I have continually benefitted from NIOSH's investment in me becoming a physician-scientist in occupational health. Whether in a clinical encounter with an exposed worker, collaborating in field studies, or slogging through the give and take of the participatory NORA process to keep the nation's occupational research agenda potent and relevant, partnering with NIOSH colleagues has been an essential and sustaining aspect of my professional life. Overall, partnering with NIOSH has made a critical difference to the safety, health, and security of the working people we serve.

Tom Leamon—While partnering with NIOSH implies making a contribution to the Institute, it also provides an opportunity to see the vital processes the Institute uses to advance its mission. The necessary compromises behind the public face—as seen on the Web site—involve matching the needs and positions of multiple stakeholders, policies, science, and budgetary considerations to provide the leadership it takes to discover and disseminate safety and health knowledge. Being a part of this process has given me a far better and more rewarding view of NIOSH and its people, many of whom I call friends. Partnering with NIOSH has also increased my commitment to improving the safety and health of Americans who go to work and deserve to return home each day uninjured and healthy.

Nicholas Mastrodicasa—NIOSH's contribution is invaluable to a joint aviation safety project between NASA, the FAA, the Medallion Foundation, and the Alaska Aviation Safety Project (State of Alaska). NIOSH provided the seed money that resulted in this project, as well as its expertise in human factors and statistical data. This project is now developing what is anticipated to be the next major breakthrough in aviation safety training products for use in aviation simulators located throughout Alaska.

Bruce Watzman—The National Mining Association has been an active participant in several research partnerships sponsored by NIOSH. These partnerships have served to bring together disparate interests to advance the understanding and development of technologies to improve miner safety and health. In so doing, NIOSH has filled a critical void to advance stakeholder understanding of the barriers that must be overcome to develop and introduce new technologies into our nation's mines.

Frank Renshaw—My perspective on partnering with NIOSH stems from serving on NIOSH's Board of Scientific Counselors, the NORA Manufacturing sector council, and years of practice in OSH. NIOSH has earned and sustained its reputation as our nation's premier OSH research and training institute. It has skillfully leveraged its intramural research through extramural grants with leading academic and other organizations. The Institute is committed to serving all working populations as evidenced by its NORA sector-approach. NIOSH is the most significant and reliable contributor to graduate-level training of OSH professionals through its network of Education and Research Centers. From its national emphasis on emerging issues, like nanotechnology, to its Research-to-Practice initiative, NIOSH is a key partner in ensuring the viability and competitiveness of America's workforce.

James Koskan—Our industry relies on the development, implementation, and monitoring of sound research-based management practices to positively impact employee well-being and lower operating expenses. NIOSH has been an ideal partner in this endeavor. NIOSH has been diligent in the design and execution of quality peer-reviewed research and has developed control tools that have given our company and our industry the confidence we need to meet safety challenges that arise.

Mei-Li Lin—NIOSH has done significant, relevant, and at times groundbreaking research that has had tremendous value in enhancing OSH. For not-for-profit safety and health organizations, like the NSC, NIOSH has been an invaluable resource. Our collaboration allows for more effective translation of evidence-based research and expanded outreach to leaders and OSH professionals. The rich and active dialogue between our organization and NIOSH has resulted in more focused, guided research and dissemination efforts that directly address industry needs. Through initiatives such as NORA, Research-to-Practice, and Prevention through Design, NIOSH continues to transform—engaging constituents and OSH professionals on a more meaningful level. NSC is very glad to be a part of these endeavors.

Peter O'Neil—Many of NIOSH's initiatives, such as Prevention through Design, directly complement, enhance, and support AIHA's strategic initiatives in the areas of information, learning, and knowledge. The NIOSH/AIHA Alliance has expanded and improved our ability to develop and disseminate information on worker safety and health through print and electronic media, has provided opportunities for participation at conferences and meetings where safety and health issues are proactively addressed, has advanced the effectiveness of occupational safety and health research, has promoted and facilitated the transfer of research results to practice to prevent work-related illness and injury, and has strengthened recruitment efforts for students to enter occupational and environmental safety and health graduate and undergraduate training programs.

Ginny Frings—I am educating students and the community on how to keep themselves, their families, their workplaces, and the environment safer and healthier through the Research-to-Practice partnership with NIOSH as well as through my roles as Professor of Business of Health and Safety at Xavier University, President of Concerned Angels, Inc., and host of the Safety Corner television show. In addition, Concerned Angels, Inc. is coordinating with the Centers for Disease Control and Prevention and hospitals in the Cincinnati, Ohio area to disseminate the new Child Passenger Safety Instructional DVD.

> *partnering with NIOSH ... has been an essential and sustaining aspect of my professional life*

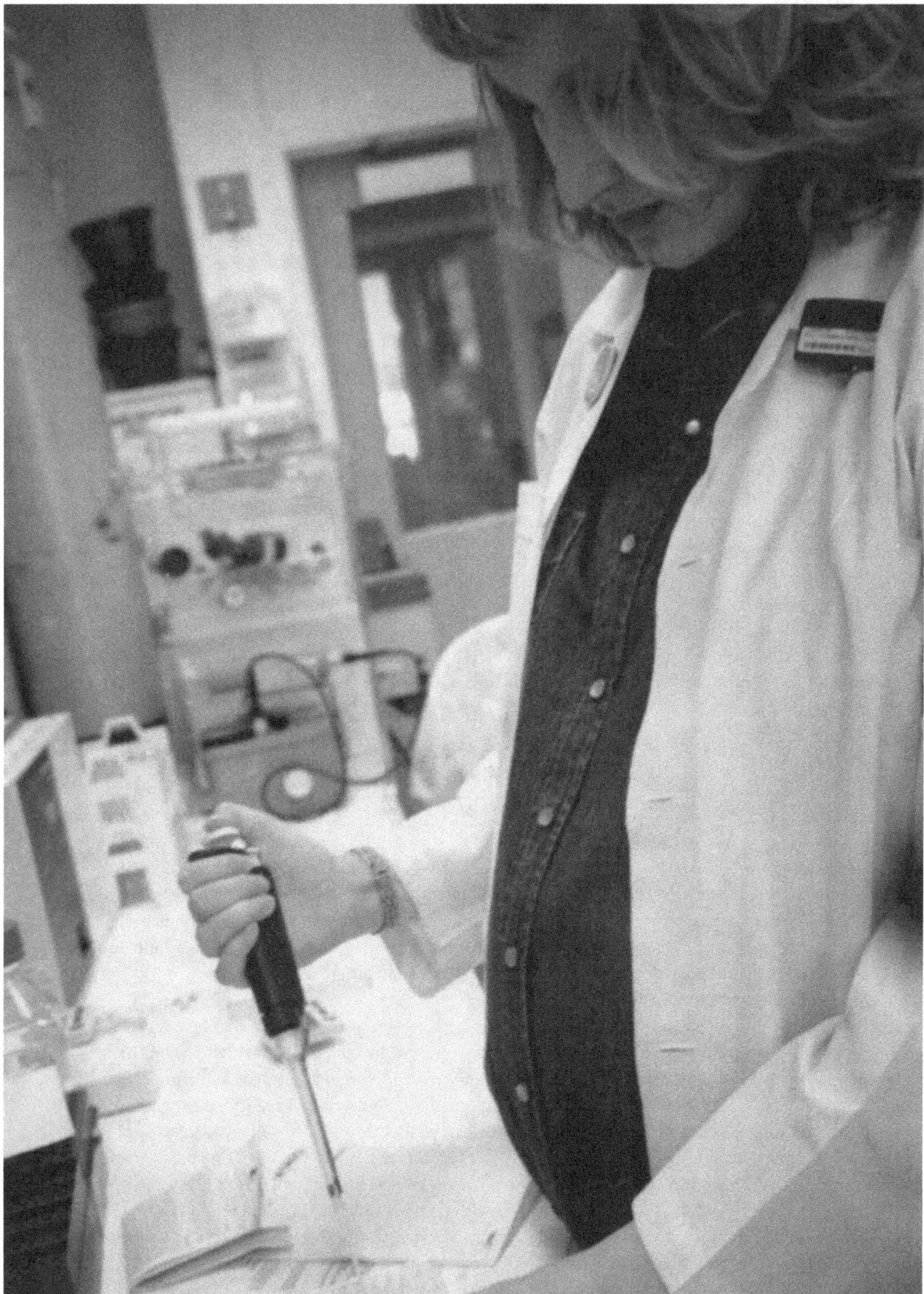
Photo by Kimberly Clough Thomas

NETWORKING BY DESIGN

Creating a Culture that Nurtures Interdisciplinary Research & Prevention Efforts

To enhance the eight NORA sector programs, NIOSH created 24 cross-sector programs organized around surveillance, hazard exposure, interventions, adverse health outcomes, information dissemination, and statutory programs. Cross-sector programs are intended to support the sectors in accomplishing their goals for the nation, coordinate priorities that affect multiple sectors, and are managed solely by NIOSH researchers through interdisciplinary steering committees.

Voluntary Initiatives Impact Safety & Health within Industry

NIOSH developed a comprehensive safety and health initiative in partnership with OSHA and the Refractory Ceramic Fibers Coalition that helped reduce occupational exposure to refractory ceramic fibers (RCF) among more than 31,000 U.S. workers. It has served as a model for protecting RCF workers in more than 12 countries. In 2006 NIOSH published a recommended exposure limit for RCF that has become the industry standard. As a result of these efforts, the RCF industry has expressed many benefits of partnering with NIOSH, stating that voluntary initiatives can be a successful and cost-effective alternative to regulation and that cooperation is greatly superior to confrontation.

ℐ Authoritative Recommendations

External stakeholders within industry, labor, professional associations, and other government agencies use NIOSH recommendations to protect workers from injury, illness, and death.

Functions of the NIOSH Authoritative Recommendations cross-sector program include scientifically assessing workplace risks and hazards; developing, disseminating, and evaluating recommendations; providing responses to other government agencies on issues of rule-making that may affect workers; and managing the NIOSH respirator certification program.

NIOSH develops several types of authoritative recommendations, the most recognized include recommended exposure limits, immediately dangerous to life or health (IDLH) values, and respirator certifications. NIOSH bases its recommendations on comprehensive evaluations of available scientific data, which may include exposure-response modeling or quantitative risk assessment. Evaluations undergo rigorous scientific peer review before being used as the basis of a NIOSH authoritative recommendation.

Authoritative recommendations are communicated through a variety of mechanisms to multiple audiences including employers, employees, policymakers, the scientific community, and the general public to prevent or reduce occupational hazards and to regulatory agencies for considering the promulgation of legal standards.

Relevant Information

NIOSH recently proposed new and improved skin notations for approximately 120 chemicals. These notations will provide workers and employers with critical information about the dermal hazards of chemicals in the workplace. The occupational safety and health community relies heavily on NIOSH skin notations to protect workers from hazards associated with dermal exposure to chemicals.

NIOSH published 15 Alerts on topics including bird flu, fire fighter fatalities, young-worker safety and health, and flavoring-induced lung disease from 2000–2008.

NIOSH is developing Recommended Exposure Limits for several high-profile occupational exposures, including diacetyl (butter flavorings), hexavalent chromium, manganese in welding fume, 1-bromopropane, pigment-grade and nano-sized titanium dioxide, and ionizing radiation.

NIOSH is developing IDLH values for > 100 chemicals identified as high-priority for emergency responders.

ℐ Cancer, Reproductive, & Cardiovascular Diseases

Approximately 40,000 new cases of cancer in the U.S. are attributable to work each year. Millions of U.S. workers are exposed to substances that have tested as carcinogens in animal studies. Less than 2 percent of chemicals in commerce have been tested for carcinogenicity.

There is significant public health concern about potential effects of workplace exposures on reproductive outcomes. Heavy metals, solvents, and pesticides are examples of substances in regular commercial use with reported reproductive or developmental effects. Estimates of adverse outcomes demonstrate the widespread impact that workplace exposure can have on reproductive health.

Carbon monoxide, nitroglycerin, and environmental tobacco smoke are other toxins encountered in the workplace that are known to affect the heart. Heart disease is the chief single cause of death in the U.S. Little is known about occupational risks for heart disease though numerous studies show a relationship between heart disease, depression, and exposure to stress at work. Other occupational exposures potentially related to cardiovascular disease include noise, shift work, and physical activity.

The chief health areas of concern for the Cancer, Reproductive, and

Cardiovascular Diseases cross-sector program are cancer, reproductive health, heart disease, and evolving areas of occupational neurological and renal disease. The program addresses these by conducting health and exposure research in working populations, providing data for the development of standards and recommendations to control health hazards, conducting basic and applied laboratory research to investigate precursors of occupational disease, developing engineering controls and interventions for workplace hazards, and developing and disseminating information to foster prevention of occupational disease.

Relevant Information

10%–20% of all recognized pregnancies in the U.S. end in miscarriage. 3% of all live births have significant malformations; 3% of these malformations are due to toxicant exposure. 7.2 million people die each year from heart disease.

¶ Communication & Information Dissemination

Despite efforts to reduce or eliminate work-related injury and illness, challenges still exist in the dissemination and use of scientific information products. NIOSH is committed to position its research activities as the most relevant and impactful source of occupational safety and health information. NIOSH communication specialists developed a set of priorities to guide this effort and ensure that NIOSH communicates the most valued and up-to-date science-based information in a timely and effective manner to protect workers.

Communication at NIOSH is carried out through its Communication and Information Dissemination (CID) cross-sector program. Working throughout and across NIOSH, CID brings together the expertise, resources, and strategies to serve the Institute's varied communication needs. This enhanced communication management approach ensures accuracy and consistency, reduces duplication, and achieves economies of scale.

CID established a core set of goals as its groundwork for tracking and servicing the Institute's communication activities. Its goals are to ensure the timely dissemination and translation of evidence-based research; support and promote the adoption of evidence-based practices; foster strategic partnerships and collaborations to disseminate and diffuse evidence-based practices; evaluate communication materials, strategies, and interventions; and advance communication research to drive the effective dissemination, diffusion, and adoption of evidence-based practices.

Well-designed and implemented health dissemination strategies supply workers and employers with the most current information to better understand health issues that could potentially impact them and the most appropriate avenues to take for direct action to protect their health.

NIOSH is committed to creating, communicating, and delivering value to our customers and encourages and assists Institute staff to employ well-designed health communication approaches to improve intervention effectiveness. This commitment drives how the Institute approaches organizational and behavior change; the products and services we market; the incentives and costs we focus on; the opportunities we select; and the places and partnerships where we interact with our audience.

NIOSH is also committed to developing new techniques for reaching potential partners. Fundamental organizational and behavioral influences offer considerable benefits to partners who actively participate in the design, implementation, and evaluation of scientific information products.

Reaching & Impacting New Audiences Through the Latest Social Media Channels

In 2008 NIOSH enhanced its leadership role within the public health community by implementing Web 2.0 applications. These applications provide opportunities for NIOSH researchers to truly engage with audiences in ways unimaginable just a few years ago. The Web 2.0 revolution is as much about changing communication culture as it is about using social networking sites, such as MySpace, Facebook, Wikipedia, Flickr, Twitter, and blogs, to reach diverse audiences with safety and health information.

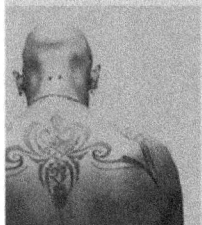

The NIOSH Science Blog—the first truly external blog at CDC—provides NIOSH with an opportunity to engage in scientific discussions with its partners and the public. The blog includes 44 unique topics and has generated more than 400 comments. Since its inception, the blog has been visited more than 100,000 times. In 2008 it averaged 156 visits per day and has risen to 245 visits per day from January 1, 2009 – May 27, 2009.

Best Practices Impact Workers & Reduce Injury Compensation Costs

Staff at nursing homes implemented best practices to prevent musculoskeletal injuries while lifting physically dependent residents. The cost effectiveness of this practice was evaluated for a 3-year period. Nursing staff experienced substantially fewer injuries from lifting residents and fewer of the actual injuries suffered required them to miss work as compared with the 3 years prior. Workers' compensation costs were also reduced as a result of these best practices. The participating nursing homes saved about $55,000 annually in workers' compensation costs, which led to the recovery of their initial investment in lifting equipment and worker training in less than 3 years.

¶ Economics

The application of economics in occupational safety and health provides a framework to identify economic inefficiencies associated with poor safety and health and points to prevention opportunities with the greatest impact.

Studying the economic conditions that influence the incidence and severity of work-related injury or illness—and the economic consequences for workers, employers, and society overall—provides guidance for optimally allocating limited resources in occupational safety and health.

Economic research also provides guidance for designing practical interventions and can underscore their cost-effectiveness. Furthermore, economics focuses on how investing in worker knowledge, skills, safety, and health influences outcomes such as productivity, earnings, and the overall welfare of employers, employees, organizations, and nations.

The NIOSH Economics cross-sector program is working diligently to advance the field of economics in occupational safety and health. The program has identified several priority areas where focused efforts are needed.

Priority research efforts include studying economic factors that may affect work and workplace characteristics and impact injury or illness rates; estimating the economic impact of injury and illness on small businesses; and understanding how injuries and illnesses that are only partially paid—or missed entirely—by workers' compensation contribute to the cost of programs, such as short-term disability and group medical insurance.

Relevant Information

Examples of economic factors that may affect the risk of work-related injury and illness include: keeping operations open **24 hours a day 7 days a week**, which requires more individuals to work on rotating or permanent shifts across these hours and can result in fatigue and lead to increased incidence of injury and illness;

Paying workers by the job or in proportion to the quantity of their work outputs, which may tempt them to cut corners on safety in order to collect a paycheck quicker; and

Using Just-in-Time or Lean Manufacturing to reduce overhead cost and improve process coordination, which can place increased stress on workers who deliver raw materials, plan production needs, or complete manufacturing tasks under strict timelines.

¶ Emergency Preparedness & Response

Police, fire fighters, and emergency medical service providers are in the top 15 occupations for risk of dying from a work-related injury.

In addition to their routine work duties, emergency responders face intense hazards from situations such as natural disasters, major hazardous materials emergencies, structural collapse, civil disturbance, bomb disposal, hostage situations, and terrorism response. Plus, the consequences of other potential hazards that have not yet been realized, such as large-scale terrorist attacks involving biological or chemical weapons, may be significant but cannot be effectively captured.

Emergency response work includes activities involving search, rescue, recovery, cleanup, and restoration. It is carried out by individuals from emergency management, fire service, law enforcement, emergency medical services, public health, construction and other skilled support, disaster relief, mental health, and volunteer organizations.

The NIOSH Emergency Preparedness and Response cross-sector program was developed to advance research and collaborations to protect the safety and health of emergency response providers and recovery workers. The current research priorities of this program are based on lessons learned and feedback received from government and nongovernment stakeholders during a series of public meetings held by NIOSH. Priority areas address safety climate, personal protective equipment, surveillance, hazard characterization, technological interventions and engineering controls, environmental microbiology, and biological monitoring of terrorism agents.

Chief objectives within each research priority have been identified to reduce injuries and enhance the safety, health, and resilience of emergency responders. Objectives include improving the organization of work; reducing exposure to inhalation and dermal hazards; developing a surveillance system to identify problems for corrective action; developing methods that emergency responders and remediation workers can use to better evaluate the spatial and temporal distribution of potential hazards; reducing various hazards to responders by improving engineering controls, technologies, and tools; improving the understanding of environmental microbiology, including factors that influence the introduction, spread, and control of threat agents; and reducing the potential impact of exposure to terror agents through improved biological monitoring methods.

Expected outcomes include generating new research ideas and hypotheses to address the leading causes of injury, illness, and death to emergency response workers; developing key research partnerships and collaborations; and increasing awareness of occupational safety and health problems in the Emergency Response and Preparedness cross-sector.

Relevant Information

Fatality rates for police, fire, and emergency medical workers are approximately **3× higher** than the average for all other occupations. In 2006, > 2 million people served in these roles.

Approximately 88,000 fire fighters are injured each year; about 2,000 of these injuries are life-threatening.

¶ Engineering Controls

Engineering controls of work-related hazards is a continually developing field due to changes in the workplace. Engineering controls are used to remove workplace hazards or lessen the negative effects of exposure to them. In recent years the emerging threat of terrorism has blurred the line between controls to protect workers and controls to protect public health.

Current research on engineering controls is being conducted in multiple NIOSH programs including Construction, Healthcare and Social Assistance, Hearing Loss, Manufacturing, Nanotechnology, Prevention through Design, and Services. Traditionally, a hierarchy of controls has been used as a means of determining how to implement feasible and effective control solutions.

Efforts of the Engineering Controls cross-sector program include promoting, planning, and conducting research on engineering control technology; providing expertise in formulating effective and credible workplace standards; and consulting in the application of effective control solutions and techniques for hazard prevention. Efforts may also include providing resources and information necessary to protect the safety and health of workers during public health emergency response activities.

Regulatory agencies, such as OSHA, use NIOSH engineering control research for their regulatory and enforcement activities.

Relevant Information

Engineering control research and use has reduced harmful exposures to **volatile compounds** during flavorings production activities, **carbon monoxide** from marine vessels, and **infectious-size-surrogate aerosols** in healthcare isolation beds by up to 99% each; **respirable dust** from jack hammer operations on concrete by up to 90%; and **asphalt fumes** during road paving by up to 80%.

¶ Exposure Assessment

The workplace can present an abundance of health hazards and exposures that can lead to illnesses and injuries that cause disability, death, and immense cost to society. Exposure assessment identifies and characterizes hazards in the workplace and prevents disease by making it possible to anticipate, recognize, evaluate, and control these hazards.

The NIOSH Exposure Assessment cross-sector program provides national and international leadership in the development and use of effective exposure assessment tools. Expanded use of these tools has led to cost-effective environmental determinations, including which workers are exposed to hazards, routes of exposure, extent, frequency, and duration of exposure, and the effectiveness of interventions.

Program priorities include fostering research and providing guidance to develop or improve exposure assessment strategies and developing or improving specific tools to assess worker exposures to critical occupational agents and stressors.

Increased visibility and coordination of exposure assessment throughout NIOSH and across a broad spectrum of government, industry, labor, academic, and international partnerships is the aim of the NIOSH Exposure Assessment program. The program accomplishes this through intramural and extramural activities, such as its initiative on direct-reading exposure-assessment methods.

Relevant Information

Exposure assessments are used by government and nongovernment agencies, including OSHA, MSHA, and NIOSH, to document the degree to which workers are exposed to health hazards, to determine what exposure levels cause adverse health effects, and to assess whether exposure levels fall within the recommended or required occupational exposure limits.

Impacting Healthcare Workers Globally

The transmission of bloodborne pathogens in healthcare workers is an important focus of NIOSH and WHO. Beginning in South Africa, Tanzania, and Vietnam, the project created a training titled, *Protecting Healthcare Workers: Preventing Needlestick Injuries Toolkit*. NIOSH, PAHO, and WHO adapted the toolkit to Latin America to strengthen educational institutions and government agencies and to conduct train-the-trainer workshops. In Venezuela the project has been expanded to 210 healthcare facilities in 12 states, the toolkit content has been included in postgraduate curricula, and 30,000 healthcare workers have been trained. In Peru more than 1,200 healthcare workers have been trained in 8 regions. Training is also occurring in Colombia, Nicaragua, and Honduras.

During the celebration of World Day for Safety and Health at Work 2009, PAHO awarded NIOSH for its leadership and contributions to occupational safety and health in the region of the Americas. This year's theme was protecting the health and safety of healthcare workers, highlighting the impact these workers have on public health.

❡ Global Collaborations

The burden of work-related injury and illness is on the rise in many countries, and the cost is immense. The ILO estimates that the total cost of global work-related injury and illness is currently about 4 percent of the world's Gross Domestic Product.

To improve global occupational safety and health, NIOSH established its Global Collaborations cross-sector program to enhance global occupational safety and health through international collaborations. These strategic collaborations with public and private-sector-based partners provide NIOSH with opportunities to better conduct innovative research, deliver expert findings, and receive new information to protect the safety and health of workers worldwide.

NIOSH provides sound scientific research, recommendations, and interventions to its partners and actively participates in the global dialogue concerning occupational safety and health issues, including silicosis, bloodborne pathogens, hearing loss, nanotechnology, radiation, and mining hazards.

NIOSH conducts activities in close partnership with several national and international organizations, including WHO, ILO, PAHO, NIH, ISO, IARC, and the U.S. State Department. Examples of recent activities include chairing the WHO Global Network of 65 Collaborating Centers on all continents to assist countries in implementing the WHO Global Plan of Action for Workers' Health; providing training in classifying chest radiographs according to the ILO Classification System; contributing to the International Program on Chemical Safety; participating in ISO and professional societies to improve the quality and dissemination of occupational safety and health guidance, research, and applications; collecting, analyzing, and sharing employer practices that reduce injuries to workers with high exposure to road traffic; and supporting and participating in partnerships between universities in the U.S. and other countries to provide opportunities for training, research, and policy development to researchers in their home countries and abroad.

The Korea Occupational Safety and Health Agency (KOSHA) recently initiated the Workplace Health Partner Program based on the experience of a KOSHA scientist who worked for a year in the NIOSH Health Hazard Evaluations program.

NIOSH is actively engaged with many countries, including Australia, Brazil, Canada, Chile, China, Columbia, Finland, Germany, India, Italy, Japan, Korea, Mexico, Netherlands, Peru, Singapore, South Africa, Sweden, Vietnam, United Kingdom, Venezuela, and Zambia.

NIOSH is committed to strengthening its current national and international partners and seeking new partnerships. NIOSH remains a leader in the global efforts to protect the safety and health of all workers.

Relevant Information

Approximately 2.2 million people die each year from work-related injury and illness. About 350,000 of these deaths result from a traumatic injury.

The global workforce of approximately 2.8 billion people suffer about 270 million serious nonfatal work-related injuries and 160 million work-related diseases each year. This takes into account nonrecorded, part-time, children, and other informal workers.

❡ Health Hazard Evaluation

Many workers are exposed to potentially harmful conditions that could affect their health. The NIOSH Health Hazard Evaluation cross-sector program offers a unique federal government resource to employers and employees in all industry sectors of our nation's economy. Occupational safety and health professionals evaluate potential health hazards in the workplace and recommend actions to eliminate those hazards and prevent adverse health outcomes.

The program uses a practical, science-based approach to solve problems in a nonadversarial environment and provides its services free of charge. The Health Hazard Evaluation program was mandated by specific provisions of the Occupational Safety and Health Act of 1970 and the Federal Mine Safety Act of 1977.

The mission of the Health Hazard Evaluation program is to respond to requests for worksite evaluations of health hazards, solve problems, communicate risk, and disseminate findings. As one of the NIOSH cross-sectors, the program has identified three priorities that drive its activities: to prevent illness through reduced exposure to workplace hazards, to promote occupational safety and health research on emerging issues, and to protect the safety and health of workers during public health emergencies.

Through this program, NIOSH determines whether workers at specific worksites are exposed to hazardous materials or harmful conditions and whether those exposures affect their health. Investigators evaluate a wide range of potential hazards including those related to chemicals, biological agents, work stress, noise, heat, radiation, and ergonomic stressors.

Health hazard evaluations can be particularly useful when workers have illnesses or experience adverse health effects from unknown causes, such as exposure to unregulated hazards; when workers experience adverse health effects even though exposure standards have not been exceeded; when a hazard is new or previously unrecognized; or when a multidisciplinary public health-oriented approach is needed.

The Health Hazard Evaluation program is carried out by occupational physicians, nurses, epidemiologists, industrial hygienists, engineers, psychologists, and statisticians. Investigators review records and conduct environmental sampling, epidemiologic surveys, and medical testing. NIOSH involves the employer and employees in the evaluation process, shares its findings with all parties, and provides recommendations to reduce hazards and improve conditions for workers.

In addition to providing direct assistance at worksites, a health hazard evaluation presents NIOSH with the opportunity to gather information on workplace exposures where occupational standards are lacking or do not protect all workers, and to share this information with other agencies, occupational health professionals, and the general public.

Relevant Information

NIOSH has completed > 13,000 Health Hazard Evaluations since 1971. They have identified emerging issues, such as a disabling lung disease from exposure to artificial butter flavorings used in popcorn and other products during production, have supported the development of occupational standards and guidance from government agencies and consensus standard organizations, and have led to the development of sampling and analytic methods.

The NIOSH Health Hazard Evaluation Program has been reviewed by the National Academies (pages 56–57).

♪ Hearing Loss

The purpose of the NIOSH Hearing Loss Prevention cross-sector program is to provide national and world leadership to reduce the prevalence of occupational hearing loss through a focused program of research and prevention activities. NIOSH is the sole

One Worker Impacting Global Change

NIOSH received a request from a worker at a large retail store concerning back injuries from changing automobile tires. In response, NIOSH contacted the company's employee safety division to learn about its tire changing process. Using the NIOSH lifting equation, NIOSH calculated the height to which cars should be raised during this process and recommended the height be increased from 18 inches to 30 inches to reduce the risk of worker injury. The company changed its policy as a result of this recommendation and eliminated the musculoskeletal hazard for workers in 7,600 stores worldwide. Musculoskeletal disorders account for 30%–40% of workplace injuries and can become debilitating. For more information about musculoskeletal disorders see page 43.

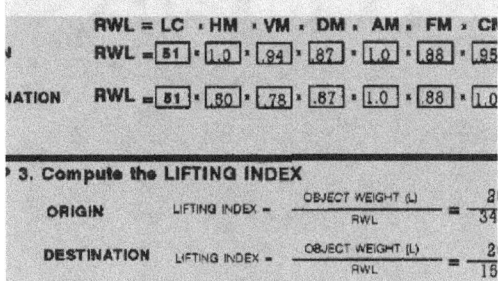

A portion of the NIOSH lifting equation.

Impacting Workers & the Bottom Line

NIOSH developed 2 engineering controls to protect underground coal mine workers from noise overexposure generated by continuous mining machine conveyor systems, which account for most overexposures. The controls are currently being produced by a manufacturer and have been implemented in several U.S. mines. As a result, sound levels have been significantly reduced, the expected life of the conveyor chain is extended because of a urethane-coated flight bar, and MSHA considers the controls "technologically and administratively achievable." Workers have said that these controls are easy to implement and highly durable. Excessive noise exposure in underground coal mines causes hearing impairment among 70%–90% of miners by retirement age.

Dual-Sprocket Driven Chain—reduces noise by maintaining constant tension & decreasing chain slack at conveyer transition points.

Urethane-Coated Flight Bar Chain—urethane coating cushions impact blows of the bar on other parts of the conveyer & reduces noise levels.

federal agency with primary responsibility for conducting research on occupational hearing loss, causal factors, and prevention methods.

This program is comprised of an interdisciplinary team of audiologists, engineers, psychologists, and physicists who actively work to reduce hearing loss—the second most reported occupational illness in the U.S. The program strives to meet its objective by conducting high-quality research, developing practical solutions to complex problems that cause occupational hearing loss, fostering partnerships with key stakeholders, and promoting the transfer and translation of research to practice.

The program currently focuses its efforts in four primary areas: prevention programs, engineering controls, hearing protection devices, and biologic and acoustic sources of hearing loss. Through targeted efforts, the Hearing Loss Prevention program is developing noise measurement instrumentation, methods, and technology; providing recommendations for noise exposure limits to form national and international policies; providing research and training opportunities; and conducting surveillance of noise hazards in the workplace.

Relevant Information

In 2008, approximately 2 million U.S. workers were exposed to noise levels at work that put them at risk of hearing loss.

In 2007, approximately 23,000 cases were reported of occupational hearing loss that was great enough to cause hearing impairment.

Reported cases of hearing loss accounted for 14% of occupational illness in 2007.

In 2007, approximately 82% of the cases involving occupational hearing loss were reported among workers in the manufacturing sector.

The NIOSH Hearing Loss Research Program has been reviewed by the National Academies (pages 56–57).

¶ Immune & Dermal Diseases

Immune and skin-related diseases represent a diverse group of health effects that are common in the workplace, including asthma, allergic rhinitis, and contact dermatitis. These and other immunologic or related illnesses result in billions of dollars in loss of workplace productivity. The greatest impact of immune and dermal disease is in the services, agriculture, construction, and manufacturing industries.

Workplace exposure to certain chemical substances can lead to contact dermatitis, allergic sensitizations, impaired immune function, including immune suppression and autoimmunity, and result in uncontrolled inflammation or increased susceptibility to diseases. Workers can be exposed to hazards in various ways including through inhalation and dermal absorption. A worker's skin can be exposed to hazardous chemicals through direct contact with contaminated surfaces, deposition of aerosols, immersion, or splashes. Skin exposures may cause systemic effects and localized skin injuries.

The mission of the NIOSH Immune and Dermal Diseases cross-sector program is to help reduce injury, illness, and death from work-related immune and skin disease. Research priorities are driven by considerations such as surveillance data, stakeholder input, and emerging issues.

Significant program activities include increasing the identification of immunologic or dermal hazards in the workplace; increasing awareness of occupational exposures to mold and other allergens; developing and publishing new information to characterize occupational chemicals or allergens that cause asthma; identifying and characterizing workplace chemicals that can cause contact or allergic dermatitis and systemic toxicity; and publishing NIOSH skin notations to alert workers, employers, risk assessors, and occupational safety and health specialists to the presence of dermal exposure hazards in the workplace.

The cross-sector program is collaborating with internal and external partners to advance understanding of occupational immune and skin-related diseases and develop strategies to prevent and control occupational exposures. It also is supporting laboratory and field investigations and promoting the development of science-based recommendations to ensure safe and healthful working conditions for all workers.

Relevant Information

Approximately 13 million U.S. workers are potentially exposed to chemicals that can be absorbed through the skin and cause adverse health impacts.

Contact dermatitis from workplace chemical exposure accounts for 10%–15% of work-related illness and costs society about $1 billion each year.

Approximately 11 million workers are exposed to at least 1 agent that is associated with work-related asthma.

Occupational factors are associated with up to 15% of disabling asthma cases in the U.S. Asthma accounts for approximately 10.1 million missed workdays for adults each year.

¶ Musculoskeletal Disorders

Musculoskeletal disorders are among the most important research issues in terms of frequency and impact across all industry sectors. These disorders are caused by sudden exertion or prolonged exposure to such physical factors as repetition, force, vibration, or awkward postures. Musculoskeletal disorders affect the body's connective tissues (e.g., muscles, nerves, tendons).

In 2007 approximately 29 percent of all workplace injuries requiring days away from work resulted from a musculoskeletal disorder; more than 335,000 cases were reported among U.S. workers in private-industry.

The NIOSH cross-sector program for musculoskeletal disorders falls under a program development activity that acknowledges the relevance of key scientific and engineering disciplines that cut across the industry sectors.

The program's mission is to reduce the burden of work-related musculoskeletal disorders through research and prevention that protects workers, helps management mitigate related risks and liabilities, and helps practitioners improve the efficacy of workplace interventions.

Two primary areas were identified for national studies. One is an ongoing national hazard survey. NIOSH started an initiative to conduct an occupational exposure survey called the National Exposure Work Survey, targeted to the healthcare industry. The study is designed for ongoing data collection in hospitals and healthcare environments regarding exposure to a wide range of occupational hazards including musculoskeletal hazards. The second area identified is more prospective where cross-sectional epidemiologic studies examine the dose-response relationship between psychosocial factors and physical loading factors, specifically the upper extremity and low back.

Relevant Information

In 2007, a total of 253,540 cases of work-related musculoskeletal disorders from overexertion were reported; 134,680 of these resulted from lifting. Another 33,960 cases of musculoskeletal disorders were reported that resulted from repetitive motion.

¶ Nanotechnology

The mission of the NIOSH Nanotechnology cross-sector program is to provide national and international leadership in investigating the implications of nanoparticles and nanomaterials for work-related injury and illness and to explore their potential applications in occupational safety and health.

Nanotechnology is the manipulation of matter on a near-atomic scale to produce structures, materials, and devices that have new or unique properties. Nanoparticles are a subset of these new materials that have at least one dimension less than 100 nanometers. Nanotechnology has the ability to transform material science used in many industries and will have numerous applications to areas ranging from medicine to manufacturing.

With any new technology the earliest and most extensive exposures to hazards are most likely to occur in the workplace. Therefore, NIOSH has engaged in a comprehensive effort to fill gaps about hazards and risks related to occupational exposure to engineered nanoparticles. This effort includes determining whether nanoparticles and nanomaterials pose risks for work-related injury and illness; conducting research and making recommendations on effective methods to manage nanomaterials safely; promoting healthy workplaces through interventions, recommendations, and capacity building; and enhancing global workplace safety and health through national and international collaborations.

By maintaining a dynamic approach involving strategic planning, research, partnerships, and information dissemination, the NIOSH Nanotechnology cross-sector program will be able to anticipate challenges and provide sound scientific recommendations for safe handling of nanomaterials.

Relevant Information

The National Science Foundation estimates that nanotechnology will have a $1 trillion impact on the global economy and employ 2 million workers by 2015—1 million of whom may be in the U.S. Worldwide funding for research and development reached $11.8 billion in 2006.

NIOSH created its Nanotechnology Research Center in 2004 to identify critical issues, create a strategic plan for investigating these issues, coordinate its nanotechnology research effort, develop partnerships, disseminate information, and accelerate progress in nanotechnology research across the Institute.

In 2006 venture capitalists invested $699 million in firms developing nanotechnology, up 10% from 2005.

¶ Occupational Health Disparities

As the U.S. workforce becomes more diverse, it has become clear that there are disparities in the burden of work-related disease, disability, and death experienced by certain groups, such as low-income and foreign-born workers.

Priority working populations have certain biologic, social, or economic characteristics that place them at higher risk of developing work-related disease and injury. These populations include racial and ethnic minorities, immigrants, younger and older workers, workers with medical or genetic susceptibility, and workers with disabilities. Occupational health disparities arise in part from an over-representation in the most hazardous industries, such as agriculture and construction. Other social, cultural, economic, and political factors including language, literacy, and marginal economic status may compound their risks. These workers may also have less access to healthcare.

Eliminating health disparities is a primary goal of the government's Healthy People 2010 objectives. Disparities are most apparent among populations with varying levels of socioeconomic status. A direct relationship exists between socioeconomic status and health, such that a person's overall health improves as their socioeconomic status improves.

Health disparities between different racial and ethnic populations are likely to be related in part to differences in socioeconomic status but may also result from other interacting factors, such as discrimination.

Theses factors are complex and still not completely understood. One area that is receiving increased attention, however, is the nature of work itself and how it can lead to social inequalities in safety and health. Research has found that low-income jobs are over-represented in occupations that may expose workers to chemical and physical hazards. In addition, the structure of these same occupations may affect workers' safety and health, such as low levels of autonomy and moderately high levels of physical demands.

The Occupational Health Disparities cross-sector program is working to identify and eliminate health disparities by promoting effective research methods and disseminating tools and information to the occupational safety and health community.

The program's priorities are to improve surveillance of excess injury, illness, and death among priority working populations by enhancing the capacity of existing systems and improving survey design and administration to consider differences in language, literacy, and culture;

to promote the elimination of occupational health disparities by expanding outreach to stakeholders, such as community-based organizations, labor unions, and government agencies, and expanding efforts to develop practical intervention programs and policies that target priority working populations;

to improve occupational safety and health research by increasing the capacity of researchers to address social and cultural dimensions of occupational health; and

to disseminate research tools and approaches that better consider factors such as language, literacy, and culture.

Relevant Information

In 2005, blue-collar jobs made up 23% of all jobs but these workers experienced 55% of all work-related injuries and earned an average of $15 per hour. White-collar workers earned more than double on average and experienced only 2% of the injuries.

In 2006, the death rate from work-related injuries for Hispanic workers was 4.6 per 100,000 workers as compared with rates of 3.8 for all workers, 3.8 for non-Hispanic white workers, and 3.9 for non-Hispanic black workers.

In 2007, workers born in Mexico accounted for 44% of foreign-born workers who died at work in the U.S.

Workers aged 65 and older are almost 3× more likely than all workers to die from a work-related injury.

¶ Personal Protective Technologies

Proper use of personal protective equipment and personal protective technologies is the last line of defense in a hierarchy of controls used to reduce worker risk of potential injury, illness, and death from hazardous workplace exposures. Reliance on the performance of these technologies multiplies in emergency and disaster situations, such as the anticipated need for adequate respiratory protection for 15 million healthcare workers in the event of pandemic influenza.

Personal protective technologies are the specialized clothing or equipment worn by individuals for protection against safety and health hazards and the technical methods, processes, techniques, tools, and materials that support their development and use. Personal protective technologies encompass equipment, such as respirators, gloves, protective eyewear, hearing protection, and protective clothing but also include face pieces, filters, guidance documents, standards, and test procedures.

In 2001 Congress allocated funds to NIOSH to develop standards and technologies for protecting America's workers who rely on personal protective equipment, with emphasis on emergency responders.

NIOSH established its National Personal Protective Technology Laboratory to provide national and world leadership for improved technologies, and to align the activities surrounding personal protective technologies occurring within its Institute. This effort was emphasized and further developed in 2005 when NIOSH established it Personal Protective Technologies cross-sector program. The program is applicable across all industry sectors and overlaps with NIOSH's Hearing Loss program, Traumatic Injury program, and Respiratory Disease Research program.

The Personal Protective Technologies program works to fulfill its mission through respirator certification as mandated in federal regulations; focused research on protection from inhalation, dermal, and injury hazards; and participation in standards-setting and policymaking.

Efforts in these areas have led to substantial accomplishments including the development of standards for respiratory protection against chemical, biological, radiological, and nuclear (CBRN) agents; adoption of the CBRN standard in Europe; adoption of the NIOSH self-contained breathing apparatus CBRN protection level by the National Fire Protection Agency; the requirement of NIOSH CBRN protection for any respirator purchased with Homeland Security funding; the demonstration of filtering face piece respirators to provide protection against nanoparticles; and mine escape strategies using new and emerging technologies. Improvements in personal protective technologies are realized through better standards and regulations and availability of personal protective equipment in compliance.

The Institute of Medicine's Committee on Personal Protective Equipment for Workplace Safety and Health (COPPE) was established in 2005 at NIOSH's request. COPPE aids the Personal Protective Technologies program in conducting quality research by providing the highest-level scientific evaluation of its projects and activities.

The program underwent a scientific evaluation recently by the National Academies in which the following was stated, "The NIOSH Personal Protective Technologies program has made effective use of its limited resources and has moved the research, standards-setting, and certification agendas forward … much still has not been done due to serious budgetary constraints. The committee offers recommendations with the goal of improving the ability of the NIOSH Personal Protective Technologies program to broaden its scope and depth of responsibility in order to protect workers more effectively…." Evaluation feedback is being used to guide the program's future priorities.

Relevant Information

Approximately 20 million workers routinely use personal protective equipment to reduce work-related exposure to inhalation, dermal, and injury hazards.

An estimated 5 million workers use respirators; approximately 20% are fire fighters.

The NIOSH Personal Protective Technologies Program has been reviewed by the National Academies (pages 56–57).

❡ Prevention Through Design

One of the best ways to prevent and control occupational injury, illness, and death is to "design out" hazards and minimize risks early in the design process. Industry stakeholders who implement prevention through design during the beginning of a project will have substantial positive impact on the safety and health of workers and likely

Assuring the Efficacy of Respirators to Impact Worker Health Globally

The NIOSH Personal Protective Technologies program is responsible for carrying out the respirator certification program. NIOSH certification provides respirator performance testing and design evaluation to assure availability of safe personal protective devices for the nation's workers. A significant outcome of the program is an increase in the inventory of NIOSH-certified respirators for protecting workers from exposure to inhalation hazards. Since 2001, NIOSH has issued more than 1,600 respirator certifications. NIOSH certification is accepted nationally and internationally as assuring the efficacy of respirators. Both OSHA and MSHA mandate that employers with workers potentially exposed to respiratory hazards use NIOSH-certified respirators.

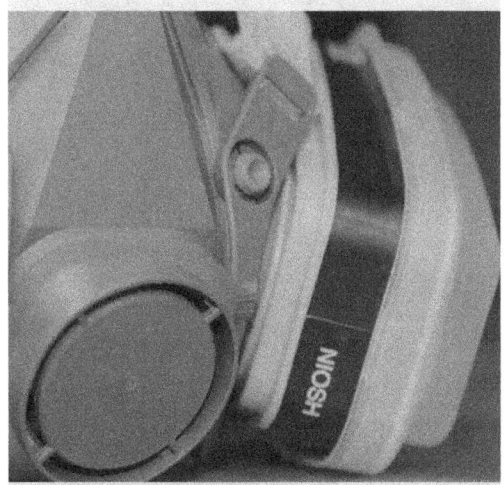

A NIOSH-certified respirator.

Research Impacts Beryllium Sensitization Among U.S. Workers

It is estimated that as many as 134,000 current U.S. workers may be exposed to beryllium. Some workers who are exposed develop sensitization to the metal and some even go on to develop chronic beryllium disease, an incurable disease that can result in serious impairment and death. Beryllium sensitization has been found in 1%–10% of workers in cross-sectional studies, with chronic beryllium disease being diagnosed in 10%–100% of those who are sensitized. Preventive efforts that relied on engineering controls to reduce air levels of beryllium have been largely ineffective. However, in 2007 NIOSH published preliminary scientific evidence that new cases of beryllium sensitization could be reduced following the implementation of a Comprehensive Preventive Program in the workplace. Beryllium sensitization rates were reduced 8-fold among workers exposed to this program as compared with those who were not exposed to the program.

A snapshot of the NIOSH Pocket Guide to Chemical Hazards.

save money rather than deal with risks inherent in completed systems.

The NIOSH Prevention through Design cross-sector program aims to achieve its mission to prevent work-related injury, illness, and death through a coordinated approach that addresses all design efforts that affect workers. This cross-sector program uses a coordinated approach that considers the unique challenges faced by businesses and the potential safety and health hazards faced by workers.

This cross-sector program established a steering committee that is made up of professionals from industry, organized labor, academia, government, and safety and health organizations. In July 2007 NIOSH held a workshop where stakeholders provided input and developed priority goals for the program. Priority areas include:

researching the engineering, economic, social, and scientific methods and implications of employing prevention through design;

increasing the knowledge and understanding of prevention through design principles among designers, engineers, safety and health professionals, business leaders, and workers;

encouraging the use of that knowledge when designing facilities, processes, equipment, tools, and organization of work;

identifying, developing, and sharing standardized processes, equipment, and tools;

persuading business leaders, workers, academics, government entities, and standards-developing and setting organizations to endorse a culture that includes prevention through design principles in all designs affecting workers; and

ensuring that small businesses have access to prevention through design resources adapted to the unique needs of their work processes and environments. The steering committee will provide leadership in executing and achieving these goals.

NIOSH is leading a national initiative called Prevention through Design to promote the concept of designing out hazards during system development and to highlight the importance of prevention through design in business decisions. Through strategic collaboration, support, and commitment of stakeholders, including prevention through design experts, associations, design engineers, business leaders, workers, safety and health experts, and finance professionals, this national initiative can save lives and protect workers. NIOSH will serve as a catalyst to establish the initiative and guide the concepts into workplaces in all eight NORA industrial sectors.

More business leaders are beginning to recognize that redesigning job tasks and work environments is a cost-effective means to enhance workplace safety and health. Many U.S. companies are supporting these concepts and developing management practices to implement them. Other countries are actively promoting prevention through design concepts too.

Relevant Information

System design is implicated in about 37% of the worker fatalities that occur each year.

Prevention through design involves evaluating potential risks associated with processes, structures, equipment, and tools. It also takes into consideration the construction, maintenance, decommissioning, and disposal or recycling of waste materials.

Design examples include lockout/tagout controls, hearing protectors, retractable needles, and latex-free gloves.

¶ Radiation Dose Evaluation

The mission of the Radiation Dose Evaluation cross-sector program is to coordinate Institute collaboration on the conduct of research related to occupational radiation exposure and to promote radiation research within the Institute to establish NIOSH as a global leader in occupational radiation exposure evaluation.

The program has developed a set of priority research goals that aim to promote the development of methods and strategies for characterizing radiation exposure in the workplace; to provide guidance to the NORA sector programs on the identification of potential radiation-related issues involving target populations; to increase cross-sector collaboration on radiation-related research within the Institute; and to effectively communicate information on radiation dose evaluation in an occupational setting to both the public and peers.

A steering committee made up of Institute experts will coordinate the development and promotion of the best available science involved in prospective and retrospective radiation dose evaluation in the workplace. The committee will determine appropriate projects and activities for this cross-sector program and participate in the execution to achieve its mission.

Relevant Information

The most recent report from the National Council on Radiation Protection and Measurements indicated that > 1.5 million workers in the U.S. were employed in occupations with potential exposure to radiation. This data is currently being updated; however, it is expected that the number has grown substantially during the past 2 decades.

Work Category	No. of Workers
Medicine	584,000
Industry	305,000
Government	204,000
Others (i.e., visitors)	155,000
Nuclear Fuel Cycle	151,000
Miscellaneous	76,000
Other Workers	15,000
Well Loggers	8,700
Industrial	6,900
U.S. Public Health Services	4,600

¶ Respiratory Diseases

Work-related respiratory diseases occur in all industry sectors and can affect any part of the respiratory tract from the nose to the deep lung. Disease onset can occur immediately following an exposure to decades later. Work-related respiratory diseases are often chronic and can lead to disability or death.

The economic burden of work-related respiratory diseases, such as asthma; chronic obstructive pulmonary disease; dust-induced diseases from asbestos, coal mine dust, and crystalline silica; respiratory tract malignancies; and respiratory infections is enormous and costs society billions of dollars each year.

The mission of the NIOSH Respiratory Diseases cross-sector program is to provide national and international leadership to prevent work-related respiratory diseases. The program fosters communication, collaboration, and coordination of efforts across the many disciplines and parts of NIOSH that work to achieve this goal. The program uses a scientific approach to gather and synthesize information, create knowledge, provide recommendations, and deliver products and services to those who can impact prevention.

In collaboration with other NIOSH programs, the Respiratory Diseases program has made essential contributions to occupational respiratory disease surveillance, basic research in determining how exposures cause disease, and a range of areas critical to disease prevention, such as epidemiology, exposure assessment, engineering controls, respiratory protection, training, and authoritative recommendations. In addition, the program provides unique national resources in the areas of lung function testing and X-ray surveillance for lung disease.

Relevant Information

About 70% of work-related disease mortality is from respiratory disease. Approximately 11.4 million adults in the U.S. have chronic obstructive pulmonary disease and about 13.8 million have asthma. Among adults 15% of both diseases is attributable to work. Dust-induced diseases, such as silicosis, coal workers' pneumoconiosis, and asbestosis, cause or contribute to > 2,500 deaths per year in the U.S.

Work-related respiratory hazards continue to emerge. Recent examples include diacetyl-containing chemical butter flavorings; flocking materials; leather conditioners; diseases caused by bioterrorism agents; and respiratory infectious diseases, such as avian influenza, subacute respiratory syndrome (SARS), and multidrug-resistant tuberculosis

The NIOSH Respiratory Diseases Research Program has been reviewed by the National Academies (pages 56–57).

¶ Small Business Assistance & Outreach

The mission of the Small Business Assistance and Outreach cross-sector program is to reduce occupational diseases, injuries, and fatalities that occur in small business establishments. This is accomplished through a comprehensive program involving research, prevention efforts, and public health activities.

The NIOSH Small Business Assistance and Outreach cross-sector program spans all workplace sectors. The program strives to fulfill its mission by conducting high-quality research; developing and disseminating practical solutions to complex workplace safety and health problems; fostering partnerships with labor, industry, government, and other stakeholders; and promoting the transfer of research to practice.

The program is prioritizing efforts to address the needs of businesses with fewer than 100 employees because these companies are assumed to have the most difficulty acquiring resources to manage workplace safety and health

Impacting Schools Through Knowledge & Guidance On Safety & Health

As part of its Safety Checklist Program, NIOSH developed more than 80 easy-to-use checklists to help teachers and administrators at high schools and technical schools establish effective occupational and environmental safety and health programs. NIOSH worked closely with professional organizations, administrators, and educators to notify schools across all 50 states about this valuable program. The program helps leaders bring their schools into compliance even when they have little safety and health experience, a busy schedule, and many unanswered questions. State departments of education and school districts in 8 states worked directly with NIOSH to get the material into classrooms and onto computer systems. Since 2003, more than 85,000 copies of this program have been disseminated in CD-ROM format. In addition, Spain's Department of Education asked NIOSH for permission to translate the program into Spanish.

issues. Workers at these small businesses also tend to suffer the highest rate of work-related injury and illness.

The Small Business Assistance and Outreach cross-sector program addresses these priority areas through collaboration with the small business community, its suppliers, service providers, and associated trade groups.

The program strives to deliver occupational safety and health information, recommendations, and new technologies in a manner that is relevant and appropriate to small businesses; engage in strategic partnerships to develop, implement, and evaluate programs that reduce workplace hazards; and demonstrate a positive return on investment for safety and health efforts within the small business community.

Relevant Information

About 90% of all businesses in the U.S. have < 100 workers. More than 5 million businesses employ < 20 workers. In the private industry, workplaces with ≤ 10 employees experience approximately 33% of all work-related deaths but employ only about 5% of the workforce.

Surveillance

Occupational safety and health surveillance is the ongoing systematic collection, analysis, interpretation, and dissemination of information used to describe workplace injury, illness, hazard, and exposure. Occupational safety and health surveillance information helps determine where work-related injuries and illnesses occur, how frequently they happen, and where scarce resources can most effectively be applied to eliminate or reduce the injury and illness burden to workers.

Whether applied to known exposures or new exposures from newly developed products, a well-designed surveillance program allows researchers to recognize potential health hazards in industry and develop interventions that will eliminate or reduce the health impact upon workers. Additionally, the reviews conducted by the National Academies of NIOSH programs in Agriculture, Construction, Hearing Loss, Mining, Personal Protective Technologies, Respiratory Diseases, and Traumatic Injury underscored the need for a robust occupational surveillance program.

Identifying and tracking occupational injuries, illnesses, and hazards for the nation is fundamental to the NIOSH mission. To this end, the Surveillance program strives to provide quantitative estimates of injuries, fatalities, and occupational diseases; detect trends, outbreaks, or epidemics of workplace injuries and diseases; monitor changes in known hazards; identify emerging hazards in the workplace; facilitate epidemiologic and laboratory research; generate new hypotheses for workplace safety and health; evaluate control and prevention measures aimed at improving safety and health; develop new technologies to improve surveillance data collection, management, and analysis; and provide results to the widest possible audience.

NIOSH actively partners with federal and state agencies and private-sector programs to conduct studies and communicate findings of occupational health surveillance research. Examples of federal partners include:

the Bureau of Labor Statistics to analyze, interpret, and disseminate data on occupational deaths from injury collected through its Census of Fatal Occupational Injuries;

the Consumer Product Safety Commission to use their existing emergency department surveillance system to capture data on occupational injuries and illnesses, including data not captured by the BLS Annual Employer Survey, such as injuries among the self-employed;

the U.S. Departments of Agriculture

and Labor to assess the burden, patterns, and trends in injury among agricultural workers and farm managers who are underrepresented in most data systems;

the National Center for Health Statistics to examine how work contributes to mortality from chronic disease using death records and assess occupational injuries and illnesses of living workers who are not captured by the BLS Annual Employer Survey;

the U.S. Department of Transportation's Federal Motor Carrier Safety Administration to track health and injuries among truck drivers;

the Environmental Protection Agency (EPA) and state partners to identify and report on work-related pesticide poisonings and broadly disseminate this information. Data from the EPA program have been used by state governments to provide broader protection to workers and community residents with potential for exposure to pesticides.

These and other NIOSH surveillance efforts are important because they routinely identify emerging issues that would otherwise be overlooked or not well communicated to employers or workers. NIOSH also provides funding to states to collect data not otherwise available at the national level and to foster data-driven prevention efforts. NIOSH supports basic occupational safety and health surveillance programs in several states and more advanced surveillance of priority conditions, including teen injuries, elevated blood lead levels, pesticide exposures, work-related asthma, silicosis, and deaths from injury.

These state-based surveillance efforts are critical for identifying emerging issues and ensuring broad-based public health activities to improve worker safety and health. For example, findings from state investigations of deaths from occupational injury of youth less than 18 years of age have been used by numerous groups to support educational efforts to reduce young worker deaths from injury and revise child labor laws.

In another example, for more than 15 years the NIOSH Adult Blood Lead Epidemiology and Surveillance (ABLES) system has worked with a growing partnership of states to systematically track laboratory reports of adult blood lead levels by industrial sector. In 2006 the manufacturing, construction, and mining sectors accounted for more than 94 percent of adults who were exposed to lead at work that had elevated blood lead levels. Industry data were submitted to the ABLES system by 35 states.

Data are used by states, industry, labor, HHS, and DOL to understand trends in workplace exposure to lead, identify where exposures occur, and evaluate how well intervention and prevention programs work. NIOSH currently supports the state-based collection and reporting of data on elevated blood lead levels in 40 states; however, state funding of basic occupational surveillance and other priority conditions is currently limited to 15 states. Many more states have expressed interest in creating occupational health surveillance programs.

Numerous partners in academic institutions, labor and advocacy organizations, government agencies, and others have joined NIOSH in addressing issues of occupational health disparities, particularly of workers with low English ability and literacy, who hold low-skill and low-wage jobs and workers who may have a mental or physical disability. Such workers occupy unique employment niches, such as home care or home healthcare for the elderly and disabled, retail, and manufacturing.

NIOSH seeks to develop methods to track and prevent workplace injury

Comprehensive Surveillance Efforts Impact Worker Safety & Health

In 2008 NIOSH surveillance of coal workers' pneumoconiosis revealed an increase in the percentage of coal workers with this preventable and devastating disease and the need for increased vigilance to ensure that exposure to coal and silica dust are appropriately controlled among miners. Other important surveillance findings recently reported by NIOSH and partners include the dramatic increase in occupational injury deaths among Hispanic workers, especially high rates of heat-related deaths among crop workers; the role that older tractors without rollover protective structures have in the large number of tractor-related deaths on farms; the large number of nonfatal injuries among workers and consumers associated with nail guns; the exposure of utility workers to home-applied bug bomb residues; and the high prevalence of work-related hearing loss in workers from a wide range of industries and occupations.

and illness and augment the limited research that has been done to assess the occupational safety and health needs of these vulnerable populations of workers.

Relevant Information

Every day in the U.S. approximately 9,300 people are treated in an emergency department for a work-related injury or illness, about 3,175 workers sustain injuries or illnesses that require days away from work, about 15 people die from a traumatic injury sustained at work, and about 135 people die from work-related diseases.

In 2006, employers paid approximately $87.6 billion in compensation as a result of work-related injuries & illnesses.

¶ Training Grants

The Occupational Safety and Health Act of 1970 mandates the development and funding of education programs to train qualified professionals and researchers to meet the need for a workforce capable of implementing safety and health programs.

Through its Training Grants cross-sector program, NIOSH funds training grants throughout the country to build a skilled workforce that can effectively implement occupational safety and health programs. According to data from Liberty Mutual, the American Society of Safety Engineers, and other organizations, reduced workers' compensation costs are directly linked to improved safety programs, including a trained workforce.

NIOSH provides support for training in several core disciplines, including industrial hygiene, occupational health nursing, occupational medicine, and occupational safety. Programs are funded through 32 Training Project Grants and 17 Education and Research Centers across the U.S. (see page 4 for locations). Both Training Project Grants and Education and Research Centers focus on providing academic and research training programs that develop practitioners and researchers who will contribute to the prevention of injury, illness, and death from exposure to safety and health hazards.

Training Project Grants are available at academic institutions that primarily have single-discipline graduate training similar to those mentioned above. The Education and Research Centers are located in institutions that emphasize interdisciplinary training and strongly encourage collaboration among the various occupational safety and health disciplines.

A few of the essential components of the centers include conducting outreach and research to practice activities with other institutions, businesses, community groups, and agencies located within their region. Also, providing continuing education programs and training courses for those involved in worker safety and health. Education and Research Centers address area needs and implement innovative strategies for meeting needs with a focus on impacting the practitioner environment.

Along with the Training Project Grants and Education and Research Centers, the NIOSH Training Grants cross-sector program supports Hazardous Substance Training programs and an Emergency Responder Training program conducted by the International Association of Fire Fighters. Hazardous Substance Training prepares professionals to properly carry out their responsibilities when involved in hazardous substance response and site remediation activities. The Emergency Responder Training also provides continuing education to help prepare responders for managing responses to hazardous substances effectively and provides training in safety and health for responders.

The NIOSH training programs have helped establish occupational safety and health programs in academic institutions across the country. It is estimated that the majority of academic and research training programs in occupational safety and health in the U.S. are supported by NIOSH funding.

Relevant Information

From 2003–2008 an average of 501 students graduated each year from a NIOSH-supported Training Projects Grant or Education and Research Center program.

About 30,000 safety and health practitioners receive continuing education training from a NIOSH-supported Education and Research Center each year.

¶ Traumatic Injury

Traumatic injuries have plagued workers around the world for centuries. Today, work-related injury and death from traumatic injuries resulting from such things as workplace violence, falls, contact with industrial objects and equipment, and motor vehicle crashes continue to claim lives, damage physical and psychological well-being, and consume the resources of workers and their families. The overall human, social, and financial toll of traumatic occupational injuries rivals the burden of health threats such as cancer and cardiovascular disease.

NIOSH has a focused program of surveillance, high-quality research, and effective translation through collaboration and practical prevention measures in order to prevent traumatic occupational injuries among the entire U.S. workforce. Research priorities of the Traumatic Injury cross-sector program are driven by available injury and fatality data. Several research and prevention efforts are under way that target sectors with high risk of traumatic injuries including agriculture, construction, and transportation.

Priority areas address increasing the use of surveillance to guide research and prevention; reducing traumatic injuries from falls, motor vehicle incidents, workplace violence, industrial machines and vehicles; and better understanding the risks and injuries occurring among vulnerable worker populations.

The program uses a traditional public health model as its research framework involving surveillance, risk factor identification, prevention strategy and technology development, intervention evaluation, and information dissemination to its audience.

Relevant Information

On average, 15 U.S. workers die each day from traumatic injuries. In 2007, > 5,650 workers died from work-related injuries, a rate of 3.8 deaths per 100,000 workers. **More than 4 million private industry workers suffered** nonfatal injuries; approximately 50% of which required days away from work, job transfer, or job restriction.

The NIOSH Traumatic Injury Research and Prevention Program has been reviewed by the National Academies (pages 56–57).

¶ WorkLife

Illnesses and injuries from work or nonwork activities reduce income, quality of life, and opportunities for both the affected workers and those dependent upon them. Preserving and improving the health and well-being of people who work is a goal shared by workers, their families, their employers, and NIOSH.

The health of workers is influenced by physical and organizational workplace exposures as well as by individual behaviors and environmental exposures at home and in the community. The effects of these various factors cannot be artificially divided between work and nonwork. Just as workplace conditions can affect health and well-being at home and in the community, exposures and activities outside of working hours can substantially affect health, productivity, and well-being while at work.

There has been a longstanding separation in the public health and occupational safety and health communities between those interested in controlling the health hazards from work and those focused on individual and community health risk reduction outside the workplace. Advancing knowledge and practice in this area requires collaboration to bridge the divide between these focus areas.

A new approach to maintaining a healthy nation of people is needed that reflects the complexity of influences on worker health and the interactions between work-based and nonwork factors. A growing body of evidence indicates that integrated approaches that address both physical and organizational health risk factors from work and individual risk factors, such as smoking and diet, are more effective in protecting and improving worker health and well-being than traditional isolated programs.

In collaboration with its partners, NIOSH developed the WorkLife Initiative to better understand and promote the kinds of work environments, programs, and policies that result in healthier more productive workers, reduced disease and injury, and lower healthcare needs and costs. It is based on a foundational commitment to workplaces free of recognized hazards and the idea that better work-based health policies and practices can help to sustain and improve the health and well-being of workers.

The premise of the WorkLife Initiative is that it makes sense to address worker health and well-being in a more comprehensive way, taking into account the physical and organizational work environment while addressing the personal health-related decisions of individuals. The worksite provides an opportunity to imple-

Trends Reinforce the Need & Potential

Healthcare costs are rising faster than wages or profits, many employers are cutting back on health benefits, and families are paying more out-of-pocket expenses for healthcare than ever before. Corporate mergers, restructuring, and economic and job insecurity are leading to increased job demands and added risks to employee health. Chronic disease rates are high and treatment is costly, highlighting the importance of effective prevention and disease management programs. The distinction between illness and injury from work and nonwork risks is rapidly fading. Work is one of the most important determinants of a person's health. Up to 70% of health determinants can be addressed in workplace programs. The U.S. workforce is aging, however, many employers are seeking ways to keep older workers on the job and continue to benefit from their experience, wisdom, and commitment.

ment programs and policies to prevent work-related risks and chronic illnesses and injuries that are linked to behavior-related choices.

Presently, NIOSH is addressing issues of balance between work and personal life more comprehensively by supporting and expanding multidisciplinary research, training, and education through the NIOSH-funded WorkLife Centers of Excellence.

NIOSH is also disseminating information about proven and promising programs, policies, and practices through new and established partnerships, including the Essential Elements of Effective Worksite Programs and Policies for Improving Worker Health and Wellbeing—practical, evidence-based guidance to help employers improve their workplace health and wellness programs.

Employers who take a comprehensive and holistic approach to worker health and well-being on and off the job enable workers to achieve a better balance between their work life and home life. They provide work environments that are safe and free of recognized health hazards, seek to reduce work-related stress, and are supportive of positive health choices for their workers.

The NIOSH WorkLife Initiative is committed to conducting and supporting research and outreach to provide employers with tools and model practices for creating a work environment that encourages worker health and productivity and helps employers promote balance between work and nonwork activities.

Relevant Information

Approximately 553,000 deaths from cancer and 1.3 million new cases of cancer occur each year. Annually, cancer costs society about $89 billion in medical costs and $130 billion for lost work days and productivity.

Diabetes-related complications lead to > 200,000 deaths, and > 23.6 million new cases of diabetes occur each year. Annually, diabetes costs society about $116 billion in medical costs and $58 billion for lost work days and productivity.

More than 80 million new cases of cardiovascular disease occur each year. Annually, cardiovascular disease costs society about $448 billion in direct and indirect costs.

Approximately 438,000 deaths from tobacco use occur each year. Annually, tobacco use costs society about $96 billion in medical costs and about $97 billion in indirect costs.

NIOSH currently supports 3 WorkLife Centers of Excellence located at 4 universities:

Center for Work, Health, & Wellbeing; Harvard School of Public Health

Center for the Promotion of Health in the New England Workplace; University of Massachusetts, Lowell and the University of Connecticut

Healthier Workforce Center for Excellence; University of Iowa

¶ Work Organization & Stress-Related Disorders

Work organization and job stress are growing concerns in all industries. Job stress results from poor work organization, such as when job demands outweigh the capabilities, resources, or needs of workers. Research has linked stress at work to lost productivity, workplace injury, and serious health outcomes, such as depression and heart disease.

A steering committee representing NIOSH divisions and offices guides the Work Organization and Stress-Related Disorders cross-sector program. The primary goal of the program is to promote and protect the safety and health of workers through research to understand work organization risk and protective factors related to stress, illness, and injury in the workplace.

The program encompasses a wide range of research, including exploring how changing organizational practices influence risk factors for job stress and other hazardous exposures; understanding how workplace stress contributes to illness, injury at work, and disease development; identifying effective multilevel intervention strategies to prevent stress at work; better understanding of the socioeconomic cost and burden of job stress; developing improved methods and tools for job stress research; and investigating stress in understudied populations.

Relevant Information

1/3 of all workers report high levels of stress.

1/2 of U.S. employers are concerned about stressful working conditions, such as long hours and doing more with less.

The WHO estimates that heart disease and depression will be the leading causes of disability by 2020; both conditions have been linked to stress at work.

Photo from the Getty Image Bank

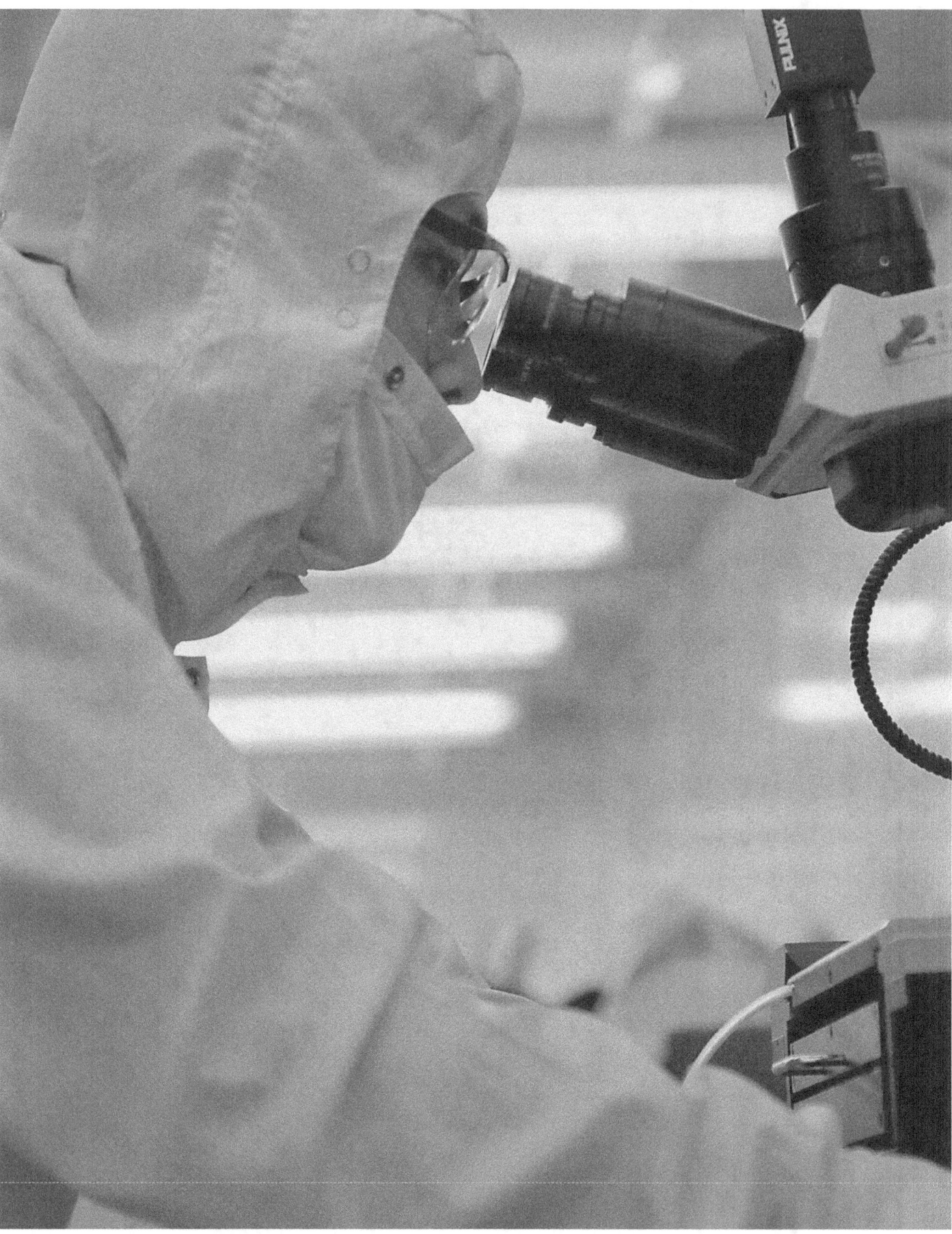

THE PIONEERING SPIRIT

What challenges you most about nanotechnology?

Four NIOSH researchers dedicated to protecting workers from the potential hazards of occupational exposure to nanomaterials share their thoughts.

Mark Methner
Industrial Hygienist

Vincent Castranova
Pulmonology Toxicologist

Vladimir Murashov
Physical Chemist

Anna Shvedova
Pulmonology Toxicologist

Several things challenge me about nanotechnology. These include keeping abreast of the constantly changing and new materials being created; identifying and deploying appropriate instrumentation to measure worker exposure; identifying appropriate exposure metrics; developing a standardized, field-deployable, sampling methodology and protocol for accurately measuring exposure; identifying appropriate analytical methods to characterize materials, process emissions, and potential worker exposure; gaining access to a variety of nanotechnology industries and materials to perform field-studies that reflect exposures across the industry; and ensuring field-study findings are communicated in a timely manner to the companies that collaborate with NIOSH.

Two of the greatest obstacles to nanotoxicology research are developing techniques to deliver nanoparticles to cellular and animal test systems and characterizing the physical properties of nanoparticles as they are delivered to these test systems. Occupational dusts are often delivered as particles suspended in an aqueous medium or by inhalation of aerosolized particles. Dispersion of fine-size particles in an aqueous medium is simple, but nano-size particles tend to agglomerate when suspended in physiological solutions resulting in structures that are no longer nano in size. Our research has shown that extensive agglomeration decreases the bioactivity of nanoparticles and leads to artifactual results. It has also received international recognition.

The greatest challenge and opportunity of nanotechnology that excites me the most is the proactive assessment and management of the occupational risks from this transformative technology. It is a challenge because it requires developing novel tools under the conditions of limited scientific data and competing priorities; it is an opportunity because such a novel occupational safety and health paradigm is aimed at prevention and will be broadly applicable to other emerging technologies. This novel approach requires global cooperation within the occupational safety and health community using existing frameworks—such as multilateral WHO, ILO, OECD, and ISO and bilateral agreements—to leverage funding, as well as human and scientific information resources.

Explosive development and application of nanomaterials are based on their unique physico-chemical properties, suggesting that their interactions with living matter may also be unique and therefore, cannot be extrapolated from previous experience. This leads us to explore unknown and unpredictable effects of nanomaterials where worker exposure may be significant. Our studies have revealed that inhalation and aspiration of single-walled carbon nanotubes cause robust pulmonary inflammation and increased sensitivity to infection. Increased understanding of mechanisms underlying health effects of nanomaterials will allow us to develop effective measures against their toxicity and scientifically sound recommendations for regulation in the workplace.

SCIENTIFIC RIGOR

Understanding the National Academies' Review

The following page shows scores awarded to each of the NIOSH research programs reviewed by the National Academies' evaluation committees. Programs received a numerical value for both relevance and impact as described in the text to the right. Each score was given on a scale from 1–5.

From 2005–2008 NIOSH had 8 of its major research programs reviewed by the National Academies: Agriculture, Forestry, and Fishing; Construction; Health Hazard Evaluations; Hearing Loss; Mining; Personal Protective Technologies; Respiratory Diseases; and Traumatic Injury. The National Academies reviewed these programs at NIOSH's request. The National Academies are known as the Advisers to the Nation and their reviews are recognized for their independence and rigor.

Within the National Academies, NIOSH program reviews were managed by the Institute of Medicine and the Division of Earth and Life Sciences. The National Academies recruited 8 separate evaluation committees; each committee was made up of top experts. In total, more than 200 scientists external to NIOSH participated in the reviews.

Before the reviews began, the National Academies convened a committee of experts to write a framework of criteria. That framework advised the evaluation committees to examine each program's inputs, activities, outputs, and outcomes; assess its relevance to important workplace safety and health needs; and evaluate its impact on improving worker safety and health. Evaluation committees were also directed to assess emerging issues for the program, provide recommendations to the programs, and score the programs on their relevance and impact.

Recommendations and noted accomplishments varied greatly between NIOSH program reviews, however, the evaluation committees did identify some commonalities. Areas for improvement included partnerships and collaborations, strategic planning, surveillance, dissemination and transfer activities, and extramural research activities. Noted accomplishments included conducting research in high-priority areas, positively affecting workforce conditions, providing excellent training programs for occupational safety and health professionals, and engaging in research translation activities.

Following the National Academies' review, each program was charged with developing an implementation plan in response to the committee's review. The plan addresses how the program intends to implement their recommendations. In addition to a public comment period, implementation plans are shared with the NIOSH Board of Scientific Counselors or the Mine Safety and Health Research Advisory Committee for review and comment and provided to the National Academies.

For more information about the National Academies' review, visit www.cdc.gov/niosh/nas/.

Agriculture, Forestry, & Fishing   Construction   Health Hazard Evaluations   Hearing Loss

Mining   Personal Protective Technologies   Respiratory Diseases   Traumatic Injury

Relevance Score Key

⑤ Research is in high-priority subject areas and NIOSH is significantly engaged in appropriate transfer activities for completed research projects/reported research results.

④ Research is in priority subject areas and NIOSH is engaged in appropriate transfer activities for completed research projects/reported research results.

③ Research is in high-priority or priority subject areas but NIOSH is not engaged in appropriate transfer activities; or research focuses on lesser priorities but NIOSH is engaged in appropriate transfer activities.

② Research program is focused on lesser priorities and NIOSH is not engaged in or planning some appropriate transfer activities.

① Research program is not focused on priorities and NIOSH is not engaged in transfer activities.

Impact Score Key

⑤ Research program has made major contributions to worker safety and health on the basis of end outcomes or well-accepted intermediate outcomes.

④ Research program has made some contributions to end outcomes or well-accepted intermediate outcomes.

③ Research program activities are ongoing and outputs are produced that are likely to result in improvements in worker safety and health (with explanation of why not rated higher). Well-accepted outcomes have not been recorded.

② Research program activities are ongoing and outputs are produced that may result in new knowledge or technology but only limited application is expected. Well-accepted outcomes have not been recorded.

① Research activities and outputs do not result in or are not likely to have any application.

Photo from the Getty Image Bank

IMPACTING DISTINCT WORKING POPULATIONS

In addition to the NORA research programs that are organized by industry sector and their companion cross-sector programs, NIOSH drives efforts to protect those who protect the public and to improve the safety and health of distinct working populations, including children, small businesses, emergency responders, energy workers, and fire fighters. This is done by leading national initiatives, funding research and intervention activities, contributing extensively to Congressionally-mandated programs, and transferring research into practice to ensure scientific relevance and impact.

Children in Agriculture

Since 1996 Congress has funded a Childhood Agricultural Injury Prevention Initiative at NIOSH, making NIOSH the leading federal agency in preventing childhood agricultural injury. The work of the Initiative focuses on developing a better understanding of circumstances that lead to youth agricultural injury and identifying effective prevention strategies through surveillance and research. Research priorities include understanding economic and social consequences of youth working on farms, developing model programs for training and supervising young agricultural workers, creating affordable and accessible childcare for children in farming communities, and establishing safe play and recreation areas on farms for resident and visiting children. In 1997 NIOSH began funding a national center for research, training, and education to prevent child agricultural injuries in the U.S.

Information Resource Impacts Farm Safety

For many producers, agricultural tourism—or agritourism—provides a way to supplement their income, entertain the public, and educate people about farming. An estimated 52,000 agritourism operations in the U.S. host more than 85 million visitors each year. In 2007 the National Children's Center for Rural and Agricultural Health and Safety published *Agritourism: Health and Safety Guidelines for Children*, a user-friendly resource for agritourism operators that includes tips on identifying and reducing hazards found on farms. In 2008 two checklists were added to this booklet to allow agritourism operators to review safety and health considerations already incorporated into the farming operation and to inform appropriate prevention and control actions.

Workplace Violence Prevention

An average of 1.7 million people were victims of violent crime while working or on duty in the U.S. each year from 1993–1999. An estimated 1.3 million (75%) of these incidents were simple assaults and an additional 19% were aggravated assaults. Of the occupations examined, police officers, corrections officers, and taxi drivers were victimized at the highest rates. As an integral part of a broad-based initiative to reduce the incidence of occupational violence in this country, NIOSH conducts, funds, and publishes research on risk factors and prevention strategies related to workplace violence.

Impacting the Safety of Small Businesses

NIOSH developed a worksite intervention program to reduce work-related injury from assault in high-risk businesses in Los Angeles. This program provided important information to small business owners who do not routinely receive such safety information. Results indicated that the intervention was effective in reducing overall crime—particularly robbery, which was the primary focus of the intervention. Several groups and organizations have received training and materials about the program to assist them in disseminating this important workplace violence prevention information. Efforts are now under way to increase the program's availability.

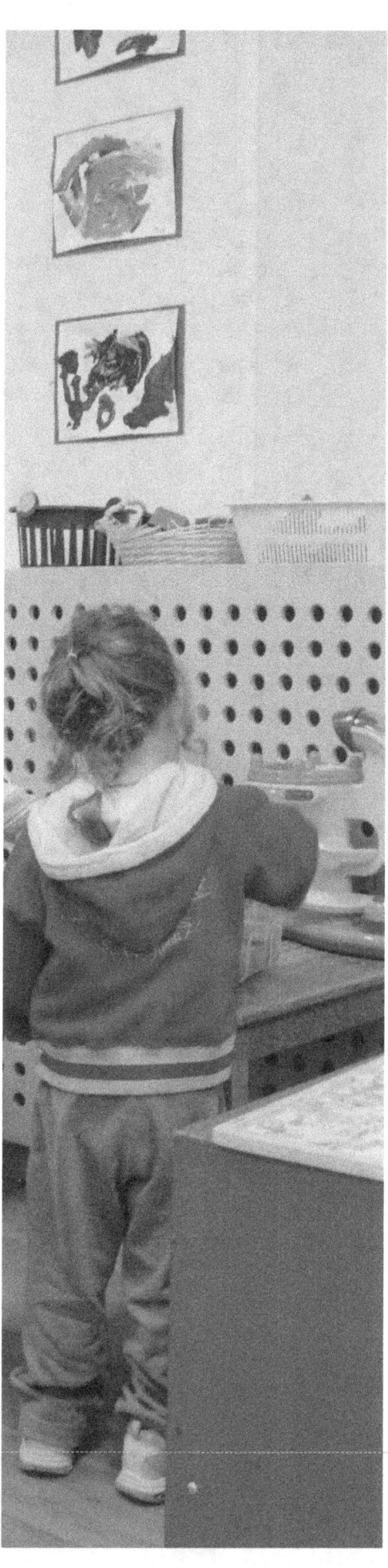

Photo by doviende, downloaded from Flickr

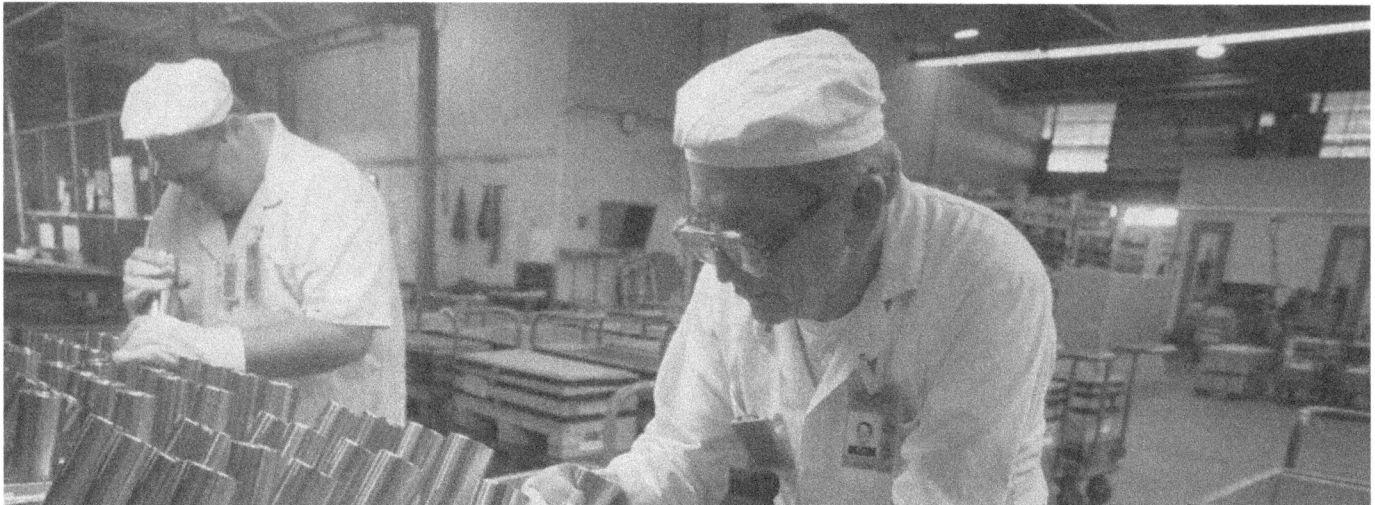

Department of Energy workers inspect uranium metal reactor fuels as part of the final product inspection process. Plant 6, Fernald Feed Materials Production Center. Photo from the U.S. Department of Energy

Emergency Responders

The World Trade Center disaster triggered a massive emergency response involving tens of thousands of emergency responders. These workers and the community were exposed to known toxins. During this disaster, NIOSH provided technical assistance to protect workers involved in the rescue and recovery efforts from hazardous exposures. Following the cleanup efforts, NIOSH now provides free medical screening, monitoring, and treatment for workers and volunteers who took part in the rescue, recovery, cleanup, and restoration activities at the World Trade Center site, as directed by Congress. Residents, students, and others in the New York City Metropolitan area are also eligible to participate. Through a consortium of grantees and contractors, NIOSH oversees programs for assessing and treating the current physical and mental health conditions appearing in these populations. Through periodic surveillance, the NIOSH World Trade Center Health Program has the ability to detect longer-term effects in a population at high risk for potentially developing multiple health conditions.

Energy Workers

NIOSH is responsible for determining occupational radiation exposure for workers with cancer who are eligible for compensation under the Energy Employees Occupational Illness Compensation Program Act of 2000. NIOSH develops and applies scientific guidelines and methods and conducts dose reconstructions to assist DOL in carrying out this Act. NIOSH's dose reconstructions are used by DOL to assist in determining the probability that a worker's cancer was "at least as likely as not" caused by workplace exposure to ionizing radiation while the worker was employed at a Department of Energy or an Atomic Weapons Employer location covered under the Act. NIOSH's complex work provides science-based methods to help carry out the law's mandate for fair, timely, and efficient compensation benefits. In addition to this responsibility, NIOSH also conducts an occupational energy research program to more fully understand cancer risk factors in workers exposed to radiation and to recommend improved protective measures to prevent injury, illness, and death.

Fire Fighters

Every year about 105 fire fighters die in the line of duty across the country. NIOSH conducts investigations of many of these deaths to develop recommended steps that the fire service can take to prevent similar deaths. NIOSH's Fire Fighter Fatality Investigation and Prevention Program has made more than 1,000 recommendations based on more than 300 investigations since its inception in 1998.

Impacting Policies & Procedures in the Fire Service

In fiscal year 2009 the Fire Fighter Fatality Investigation and Prevention Program produced its own "greatest hits" document titled, *NIOSH Fire Fighter Fatality Investigation and Prevention Program: Leading Recommendations for Preventing Fire Fighter Fatalities, 1998–2005*. This document summarizes the most frequent recommendations from investigations conducted during the first 8 years of the program. The document was developed to provide fire departments with relevant information that they can use to develop, update, and implement policies, procedures, and trainings to prevent injury and death among fire fighters. The document is available at www.cdc.gov/niosh/docs/2009-100/pdfs.

Driving Change Through Research Translation

Building a solid connection between scientific results and their use is an essential component of high-quality research organizations. In the mid-1990s, NIOSH became keenly aware of the need to more effectively demonstrate the value of its research. Workers, employers, policymakers, and industry representatives began to work more collaboratively to improve the safety and health impact of the Institute's research.

In response to our partners' perceptions, NIOSH began transitioning into a more impact-focused agency. We involved partners and stakeholders more extensively in setting research priorities; partnered with industry, academia, professional associations, labor representatives, and government and nongovernment organizations to accomplish priority goals; and created an office devoted specifically to improving the relevance of our science and increasing our impact on workers and workplaces. Relevance and impact—along with scientific credibility, quality, and integrity—are of utmost importance to the Institute.

As a way to strategically conduct and support relevant and impactful research, NIOSH launched its Research-to-Practice (r2p) initiative in 2003. r2p focuses on translating and transferring NIOSH research findings, technologies, and information into effective injury and illness prevention practices used in the workplace; emphasizes that research conducted or funded by NIOSH should be relevant for recognized evidence-based policy and practice in the field of occupational safety and health; effectively diffuses research outputs to individuals or groups most likely to adopt them; and ensures that outputs are usable by organizations with the power to improve workplace safety and health. Opinion leaders and innovators must be familiar with NIOSH research and understand its utility if we are to have confidence that our research findings and best practices will have impact and prevent work-related injury and illness.

As part of its r2p initiative, NIOSH is both demonstrating and increasing its contribution to science and occupational safety and health practice to maximize return on the nation's investment. The Institute is collaborating with multiple partners to develop effective prevention products, translate research findings into practical and understandable information, guide tailored dissemination efforts to workers and others, and evaluate its effectiveness. Transparency and accountability is crucial for research organizations such as NIOSH, especially in a period of diminishing federal resources.

To ensure scientific relevance and impact, NIOSH is committed to engaging partners and stakeholders throughout the entire research process and continuing its comprehensive efforts in r2p. Also, it is essential that both internal and external scientists, decision makers and policymakers, and other key stakeholders understand and contribute to the mission of NIOSH to generate new knowledge in the field of occupational safety and health and to transfer that knowledge into workplace practice to prevent work-related injury, illness, and death.

Design Innovation Impacts Bicycle Officers

Traditional bicycle seats (saddles) have been associated with urogenital problems (paresthesia) and sexual dysfunction in bicycle police officers. In a recent study, NIOSH provided nontraditional no-nose saddles to bicycle officers from 5 metropolitan areas to use over the course of 6 months. NIOSH investigators instituted a series of measures to identify improvements that occurred from using the saddles. Results showed a 66% reduction in saddle contact pressure and significant improvements in penis tactile sensation and erectile function. The percentage of men who reported not experiencing urogenital paresthesia while using the no-nose saddle for 6 months rose from 27% to 82%. Of 90 officers assessed, only 3 returned to a traditional saddle. One police department from the study ordered 400 no-nose saddles. The results of this research are having a demonstrable impact on 40,000 bicycling police officers and an indirect impact on the 5 million recreational bicyclists in the U.S. as seen in many blog discussions.

Photo by Kitseeborg, downloaded from Flickr

Research & Application Awards

In fiscal year 2008, more than 185 NIOSH scientists were recognized for their scientific excellence; innovations; instructional, technical, and educational materials; collaborative partnerships; effective application of research; and leadership in advancing the field of occupational safety and health.

Alice Hamilton Science Awards

Educational Materials
James Albers
Cheryl Estill

Biological Sciences
Victor Johnson
Berran Yucesoy
Jeff Reynolds
Kara Fluharty
Wei Wang
Diana Richardson
Michael Luster
Honorable Mention:
James Antonini
Sam Stone
Jenny Roberts
Teh-Hsun [Bean] Chen
Diane Schwegler-Berry
Aliakbar Afshari
David Frazer

Engineering & Physical Sciences
R. Karl Zipf Jr.
Michael Sapko
Jürgen Brune

Human Studies
Misty Hein
Leslie Stayner
Everett [Chip] Lehman
John Dement
Eileen Kuempel
Ralph Zumwalde
Randall Smith
Dana Loomis
Steve Gilbert

Bullard-Sherwood Research-to-Practice Awards

Knowledge
Kathleen Kreiss
Richard Kanwal
Greg Kullman
Lauralynn Taylor McKernan
Kevin H. Dunn
Ann Hubbs
Robert Streicher
Andrea Okun
Stacey Anderson
Jeffrey Fedan
Heinz Ahlers
Nancy Sahakian
Rachel Bailey
Chris Piacitelli
Randy Boylstein
Kathleen Fedan
Brian Tift
James Couch
William [Travis] Goldsmith
David Frazer

David Weissman
Stephanie Pendergrass
Laurence Reed
Virginia Sublet
Vincent Castranova
Dale Porter
Lori Battelli
Diane Schwegler-Berry
Robert Mercer
Michael Kashon
John Wells
Adam Fedorowicz
Gregory Wagner
Leon Butterworth
Barbara Meade
Albert Munson
Janet Dowdy
Alan Echt
Alberto Garcia
Laura Hodson
Ardith Grote
Kenneth Brown
Charles Neumeister
Patricia Schleiff
Joe Burkhart
Robert Castellan
Christopher Coffey
Deanna Cress
Jean Cox-Ganser
Gregory Day
Nicole Edwards
Susan Englehart
Diana Freeland
Steven Game
Eva Hnizdo
Mark Hoover
Denise Gaughan
Thomas Jefferson
Michele Tennant
Raymond Petsko
Eileen Hayes
Michael Beaty
Toni Bledsoe
Sherri Friend
Dean Newcomer
Amar Mehta
Jennifer Keller
Robert Lawrence
Stephen Martin
Edward Petsonk
Terry Rooney
David Spainhour
James Taylor
Mei Lin Wang
Sandra White
Christie Wolfe
Muge Akpinar-Elci
Omur Cinar Elci
Paul Enright
Ahmed Gomaa
Kevin L. Dunn
Amber Harton
Kenneth Hilsbos
William Jones
Barbara Bonnett

Dan Yereb
Kimberly Stemple
Jeffrey Reynolds

Intervention
Steven Schrader
Michael Breitenstein
Brian Lowe
Honorable Mention:
Timothy Merinar
Frank Washenitz

Technology
Pengfei Gao
Tyson Weise
Beth Tomasovic
Honorable Mention:
Jennifer Lincoln
Robert McKibbin
Chelsea Woodward
Devin Lucas
John Bevan
Eric Esswein
Mark Boeniger
Kevin Ashley

James P. Keogh Award
Mitch Singal

Charles C. Shepard Science Award for Lifetime Scientific Achievement
Vincent Castranova

ASHP Award
Thomas Connor

NHCA Outstanding Hearing Conservationist Award
Thais Morata

Society of Risk Analysis Best Paper Award in Health Sciences
Matt Wheeler
John Bailer

The Ergo Cup®
James Albers (finalist)
Cheryl Estill (finalist)

Josef Hoog Award
David Byrne

National Academy of Engineers
William Hustrulid

Edward J. Baier Technical Achievement Award
Kathleen Kreiss

Meritorious Service Medal
Mark Toraason

Outstanding Service Medal
M. Eileen Birch
Lauralynn Taylor McKernan
Bruce Newton
Gregory Piacitelli
Laurie Piacitelli

Dr. Marvin Mills Award
James Kesner

CDC Engineer of the Year Award
Bon-Ki Ku

ASTM International President's Leadership Award
Angie Shepherd

NORA Photo Contest
Robert McKibbin
Belinda Johnson
Martin Harper
Anita Wolfe
Brad Husberg
Kimberly Clough Thomas
Best in Show:
Patrick Dempsey

NORA Partnering Awards
Nancy Nivison Menzel
Nancy Hughes
Audrey Nelson
Tom Waters
Kathleen Kreiss
Rachel Bailey
Kristin Cummings
Gregory Day
Paul Henneberger
Mark Hoover
Ann Hubbs
Erin McCanlies
Christine Schuler
Marcia Stanton
Aleksandr Stefaniak
Carrie Thomas
Brian Tift
M. Abbas Virji
Ainsley Weston
Brush Wellman Inc.
Brush Ceramic Products Inc.

Award of Excellence for Public Health Training
Nancy Menzel
Nancy Hughes
Audrey Nelson
Tom Waters

FLC Midwest Region Excellence in Technology Transfer Award
Kevin Ashley
Eric Esswein
Mark Boeniger

CDC Annual Photo Contest
Sung-Chul Seo
Anita Wolfe
Gregory Molinda
Tom Bobick

Director's Award for Scientific Acheivement in Occupational Safety & Health
Paul Leigh

NORA Innovative Research Award
William Shaw
Michelle Robertson
Santosh Verma
Robert McLellan
Glenn Pransky
Ronald Woo
Mary Jane Woiszwillo
Honorable Mention:
Thomas Bobick
EA McKenzie Jr.
Douglas Cantis
H. David Edgell

CDC Director's Innovation Award
Jennifer Lincoln
Robert McKibbin
Chelsea Woodward
Devin Lucas
John Bevan
Timothy Merinar
Frank Washenitz
Barry Newton
Pengfei Gao
Tyson Weise
Beth Tomasovic

Excellence in Aerosol Research Award
Paul Baron

Performance America Consortium Award
HETAB

EHOPAC's Edward [Ted] Moran Award
Lauralynn Taylor McKernan